A LEADER'S JOURNEY TO QUALITY

A LEADER'S JOURNEY TO QUALITY

DANA M. COUND

ASQC Quality Press   **Milwaukee**

Library of Congress Cataloging-in-Publication Data

Cound, Dana M.
    A leader's journey to quality / Dana M. Cound.
      p. cm.
    Includes index.
    ISBN 0-87389-211-9 (pbk.)
    1. Total quality management. I. Title.
[HD62. 15.C68 1992]
658.5'62—dc20   92-18924
                                                        CIP

This book is printed on recycled paper.

ASQC Quality Press
611 East Wisconsin Avenue, Milwaukee, Wisconsin 53202

Current printing (last digit):
10 9 8 7 6 5 4 3 2 1

PRINTED IN THE UNITED STATES OF AMERICA

To my other half, Bonnie; my wife of forty years
So many moons in the same canoe

Preface

The achievement of quality requires more than just the skillful application of technology and management science. Most of us have had occasion to witness massive training programs that resulted in no change of behavior within the organization. We have taught valuable skills, statistical process control for example, only to see them applied indifferently or not at all.

We have put in place organizations and staffed them with people with all the requisite skills and certifications, organized them into departments in accordance with a grand scheme recommended by a noted authority—only to watch in frustration as they failed to deliver their promise.

We have bought books by the thousands, by the tens of thousands, and distributed them widely within our organizations to managers and thought leaders. We have endorsed their contents and indicated that these would become the guiding principles for our organizations, only to watch the books go unread or the principles mouthed but not practiced.

We have produced policies, mission statements, lists of values, yet little in the behavior of our organization reflects these policies and values.

No, management by cookbook has demonstrated little success—and cookbook may be a very apt analogy. A cookbook is a device with which to package a winning formula so that it may be transported to other times and places and replicated there in someone else's kitchen. Cookbooks define ingredients, proportions, and basic requirements for their combination, processing, and delivery—seemingly everything required to reproduce a world-class performance. Yet that recipe in the hands of a hundred cooks will produce a bell-shaped distribution of results, ranging from a few notable successes to a few near poisonings with a massive amount of indifference in between. What is missing that defies packaging is the skill, sensitivity, experience, aptitude, and focused passion of the cook.

Let's examine the birth of a recipe. Someone experiences an outstanding culinary performance, observes the process in action, and documents the observable. Now put that recipe into the hands of an uncommitted amateur and put the original cook at his elbow while the amateur renders the recipe. Prepare to watch the total disintegration of a fine cook. There will be no aspect of the performance that will not offend the guest expert. Some important elements are simply beyond documentation, or perhaps, as a professor the author once had would say every time a vexing question was asked, "That question is beyond the scope of this course."

Similarly, in the management of quality someone observes an outstanding performance and documents the observable ingredients, the proportions, and the order of introduction. He watches the process work and the results come forth. The winning formula is promulgated for all to emulate. The result is the same as with the recipe: few outstanding performances, a few disasters, and a lot of indifferent results in between until enough evidence piles up to convince the expectant world that we have before us another stillborn promise of the winning formula. Nor is *this* the panacea.

Was the formula bad? No. Was the performance to be emulated not world-class after all? No. It's just that something very ethereal yet absolutely essential was not observed, was not documented, or worse yet was presumed to be universally present and therefore not of consequence in the original performance.

That indispensable ingredient is leadership! Passionate, compelling, unrelenting leadership. A leadership so transparent that its every bit of body language, its every statement, its every decision is consistent and compelling. A leadership that can explain itself in terms that all can understand and most can accept. And that leads us to another set of observables.

We have seen organizations equipped with the latest tools and technology, staffed with enviable numbers of highly qualified and committed people, well managed but poorly led, fail. We have also seen organizations lacking all these things struggle and stumble and persevere and succeed. This is not to say that the technology, tools, and organizational skills are not important. They absolutely are and the struggling, successful organization will ultimately discover them. The point is that the catalyst that brings to life the inanimate structure of technology and organizational skills is leadership.

This book, then, is about leadership, leadership that evolves and grows and finds its way. It is about a melding of technology and organization with purpose and a growing sense of mission. It is about a journey, not a destination. Its ambition, a big one, is to put in context all the ingredients essential to success. *The* formula? No! Too many formulas have worked for anyone to hold that one particular formula is the only true way.

But there are issues to be faced that are common. There are stages that all must pass through. There are critical junctions that must be encountered with the right decisions made. And there are also errors that must be made. To a very great extent, this is a journey on which every person is his own Columbus.

This book will take you on a journey. Not necessarily your journey, for that you must take by yourself, but on another man's journey. As you parallel him you will have one track in the right brain of the would-be leader. The other will be planted in the left-brained world of technology and "management science" (if that's not a contradiction in terms). The device to accomplish this will be the journal of the executive. He has no name, neither does his company. We know little of it but it appears to be large. He will discover, he will make watershed decisions. He will fail, like you and me. He will also at times be funny, at times tragic, but mostly ordinary. But "folded into" his story along with his journal will be a lot of left-brained information of the type that he should have as he takes his journey, information about things that work and things that don't, and why.

If you are looking for a cookbook, look elsewhere. This book is more about cooking than it is about recipes.

Also, let me apologize at the outset, lest I offend anyone's legitimate sensitivities: our hypothetical leader is a man. Not all leaders are men. I will also find at times that the terms "man" and "men" lend themselves better to the style of expression than more "sex neutral" terms. I hope you will indulge me and accept this as a matter of style, not an expression of sociological bias.

Dana M. Cound

Acknowledgments

In one sense this book was written over a period of a few months of part-time activity. In a more real sense it was written over a period of 35 years. Putting it on paper has caused me to realize that very consciously. It is fitting, therefore, that I acknowledge as a group those who contributed to its writing by letting me observe as they struggled in the arena, made mistakes, and, in most cases, triumphed. But not all triumphed. I also owe a debt to those who didn't belong in the arena, some because of aptitude and others because of character. They taught me how important both those virtues are.

The greatest debt is due to those who stood by supportively as I made my mistakes and struggled in my corner of an arena for which they were responsible. I am thinking particularly of two people. One was my first boss, Jim Thresh, who tried to teach me that "quality is people." The idea didn't seem very profound at the time. It makes a lot of sense now. The other was my second boss, Tom McDermott. Tom assured me, just as I was about to give him a decision on his job offer, that "I'm an SOB to work for." Well, he was. Moreover, he was one of the most wonderful SOBs anyone *ever* worked for. I am still learning some of the lessons that Tom gave me. I am fortunate that each was my boss at the right time in my career. Most of all, I am grateful that both are my friends today.

As far as getting this on paper is concerned, there are some other people to be recognized. First is Jeanine Lau, Acquisitions Editor for ASQC Quality Press. It was her gentle, but persistent nudging over a period of several years that finally moved me to the point of making the commitment.

Finally, there is my severest and most valued editor, my wife of 40 years, Bonnie. Her unshakable confidence and optimism, her gentle but firm advice, her time, her support, and her sharp pencil made this book possible.

Thank you all.

Contents

PART III: ECONOMICS OF QUALITY

PART IV: STATISTICS AND OTHER LAWS OF NATURE

Introduction

The pursuit of excellence is probably the noblest ambition in all of business. It focuses on the needs of others. It sets high and enriching standards for the entire organization. It bonds all the players together in pursuit of worthy, common objectives. It is also true that the pursuit of excellence in any meaningful way is rare in today's business world driven as it is by short-term objectives, demands of third parties purporting to represent the interests of employees, stockholders, communities, various citizen groups, and a host of others. Yet all these interests fit within the context of pursuit of total organizational excellence.

On the other hand, when the word "quality" is used, most of us immediately focus on the products of the organization. If we are truly broad in our thinking, we may also think of its services, particularly if we find ourselves in one of the "service industries." But all industries are service industries. The products of the most traditional manufacturing company are purchased because of the services those products offer the consumer. The physical product is simply the package in which the service is delivered.

The word "excellence" seems to apply better than "quality" in this line of thought. It encompasses more comfortably the notion of service quality as well as product quality. It suggests superior quality as opposed to competitive quality. It calls into question the quality of the processes that produce those products and services as well as the products and services themselves. It speaks of the quality of the people of the organization and how they work together. It reminds us that superior quality products and services are rarely, if ever, produced by inferior organizations. You simply don't get good fruit from a bad tree.

But rarely do companies start out to achieve total organizational excellence. They begin on a humbler journey. Their products and services have been weighed in the marketplace and found wanting. In confronting the challenges of survival they learn about themselves, their weaknesses but also their strengths, their limitations but also their potential. What begins as a quest for competitiveness matures into an unquenchable thirst for excellence in everything they do. We will begin where most do—on a journey to quality.

This book deals in large part with the "soft side" of achieving excellence. I use that term even though it annoys me to do so. The issues to be developed and the skills to be mastered are difficult. Entering this dimension of the organization takes courage. In a sense it's the attic of the organization, full of cobwebs, artifacts from preceding generations, scurrying creatures, smells, and things that seem to go bump in the night. It is the centroid of the organization's culture. Looking squarely at it, ventilating it, and accepting the challenge to lead it to change is *not* a soft issue. But I use the term because it suggests to most of us skills and facets of performance that are elusive and that do not lend themselves to codification and reduction to formula. All of that is true. It also suggests to some the less relevant skills, the secondary issues, the venue of wimps. Don't make that mistake.

This book also deals with the "hard skills" involved in achieving quality. These subjects are well represented in the literature of the field, so they will not be treated exhaustively. They will be dealt with conceptually, and in the context of a journey to be taken.

PART I

Leadership and the Organization

CHAPTER 1

The Paradigm Paradox

JOURNAL: Yesterday was the worst day of my professional life—or maybe the best, only time will tell. New Year's Day. Last year's journal closed, next year's not yet open. As has become my custom, while waiting for the bowl games to begin, I reviewed the year just passed in the pages of that journal.

A good year begun on a positive note. Plenty of promise, a good team, goodwill among its members, a respected company, that sort of thing. But as I read through the journal with its record of bumps and bruises, minor problems, and successes, all of which seemed so important at the time, a feeling of unrest began to grow within me. It all seemed so familiar. Then it began to dawn. The first entry in last year's journal was almost identical to the last. For all of our ''busyness,'' where had we gone? Had we somehow confused motion with progress, the expenditure of energy with doing work? My brave speeches to the staff, my pointings with pride, my viewings with alarm suddenly all seemed so hollow and pointless. Why?

My mind scrolled back to the annual meeting of our industry and its ultimate self-indulgence, the annual awards banquet. I proceeded to have sort of an out-of-body experience. I saw the head table, a seemingly endless parade of distinguished leaders smiling amiably and benignly at each other while with all assumed modesty accepting the obeisance of the assembled multitude. The speeches, the Man of the Year Award. Let's see, whose turn was it this year? All those tables. Then the fog of euphoria began to clear. Half the people at each table were suppliers. The other half were purchasing agents and executives of the customer companies. Was this nothing more than a bacchanal, with a little business thrown in? A time of respite from the reality of what is happening to us and to our industry? And there I was, sucking it up as greedily as the next guy.

This was right out of the Christmas Carol. The ghost of Christmas past. Quick, the previous journal, and the one before that! All the same. How great to have been the captain of the boat, shouting orders to the crew, and how stupid not to realize that we were in the hands of a cruel current sweeping us to the precipice of a waterfall that would not respond to my orders.

Back to the banquet. Where were the faces, the accents of the real leaders of our industry today? Not at THAT banquet. While we nostalgically celebrated our product triumphs of the '50s and the '60s they were designing and delivering the products of the '80s and the '90s. We were complacently rooted in the past. They were planted in the future.

Then the ghost of Christmas yet to come arrived. In a few years, a very few, many of us at that table would be gone. We, into comfortable retirement, still attending banquets but the companies that we have been the stewards of, gone too. Replaced by other people from all over the globe, speaking other languages. Our factories, our workers wearing other insignia, extending their obeisances to other managements. And we would be among the first to go. I could be the last head of this great company. HELL, we helped create the DAMNED industry!

There are choices. One is to ignore it, smile a lot, attend the banquets, retire, and let the next guy worry about it. Or perhaps prepare to quietly die as a company—a graceful, low-profile exit. There is another alternative. . . . Remember the American officer during the Battle of the Bulge near the end of World War II. McAuliff was his name, I think. He was in a similar situation. The Germans had him surrounded at Bastogne. They gave him the opportunity to surrender gracefully. His answer? Nuts!

What will it be? Retire? Die? Or nuts? "Nuts" means more than a battle. It's the whole war. It means sticking my personal neck out. Once I call attention to the problem it becomes mine to solve. Then there's the old joke "How many psychiatrists does it take to change a light bulb?" Answer—"It depends on how much the light bulb wants to change." Well, how many CEO's does it take to change an *organization that probably doesn't **want** to change? Beats me! But we've only got one, and I'm it. Have I got what it takes?*

  Retire, and I'll never know. . . .
  No company dies gracefully. . . .
  Nuts! . . . Now, where do I start?

The Power of Paradigms

Paradigms Explored

According to the *American Heritage Dictionary*, the first definition of the word "paradigm" has to do with grammar. The second definition is "an example or model." It's in this sense that we are interested in the idea.

We all have in our mind a model of reality. Now it's not essential for the model to reflect the truth in all respects, but there is no question that the model affects our behavior in the most fundamental ways. This model or paradigm filters all the information presented to us and helps us make some fundamental decisions, such as whether or not to accept the information, for one, and how to act on it should we accept it, for another.

When someone presents us with some information or a conclusion based on observation, we take it in, mull it over for a brief period, and perhaps respond, "That makes sense." Or perhaps, "That doesn't make sense." What we have done during those few moments is to compare the information we have been given with our model of reality to see if they are compatible. If they are, a "that makes sense" answer emerges. Note that in the final analysis the data itself has little to do with our acceptance or rejection. It can be objective, scientifically accumulated, elegantly organized, and reduced—in short, irrefutable! But if the data doesn't fit our paradigm, it will usually be rejected. If we don't reject it outright, it will at least take Herculean discipline on our part to entertain it. On the other hand, the silliest of old wives' tales will be routinely accepted without question if it comfortably fits our paradigms.

But paradigms can be even more insidious than this. There is a mountain of experimental evidence that paradigms even filter what we perceive. Trick card decks laced with black hearts and red spades can be used to demonstrate that people whose paradigms include red hearts, not black, and black spades, not red, won't even see cards that do not fit this model when they're flashed before them, even slowly.

Noted authority on paradigms Joel Barker relates the following true story. (Note that amusing anecdotes are invariably about things that happened to the other guy.)

The Swiss watch industry in 1968 accounted for 65% of the world market in personal timepieces, watches. According to some industry experts, they accounted for over 80% of the profits. A scant 10 years later they had less than 10% of the market and were in the process of laying off 50,000 of the 65,000 watchmakers in their employ. What accounts for what may have been the most profound swing in market domination in business history? The quartz crystal watch movement, of course. Care to guess who invented the quartz crystal movement? That's right, the Swiss, in their own Neuchatel Research Laboratories. When the quartz crystal watch was presented to the watchmakers in 1967, they dismissed it out of hand. An interesting toy, perhaps, but not a watch. After all, it had no mainspring! And it had none of the parts that make a watch a watch. They were so sure of their paradigm of what a watch was that they didn't even protect the idea. Their research people exhibited it at their annual watch congress, where it was seen by Texas Instruments and Seiko, who clearly were not hung up on the Swiss's paradigms. The rest is history.

Yet, our paradigms are absolutely essential to life in society as we know it. In fact, a good layman's definition of a sane person might be one whose paradigms reflect a commonly held model of reality. But they can also render the truth invisible or cause us to reject objective truth even when it is standing right before us.

About Paradigms

A logical question is "How does one go about changing his paradigms?" This is an excellent and important question. But to start at the beginning, let's first consider where paradigms come from; do we carry about one paradigm or many, what are their sources, how do they relate, how are conflicts between paradigms resolved?

In a macro sense each of us has a single paradigm. It is our view of how the world works and our role in it. But that macro paradigm is composed of a number of smaller paradigms. They govern our perceptions and actions in the various sectors of our lives. Most of these paradigms overlap one or more of our other paradigms, but they do not necessarily overlap all our other paradigms. That there are interactions among them is beyond argument.

Our paradigms have a number of sources. One certainly is the knowledge we acquired from our parents. Whether accurate or inaccurate, we heard a lot about not going swimming for a half-hour after eating, what good boys and girls do and don't do, the evils that will befall us if we don't wash our hands before eating, the necessity for thrift and not wasting food given the undernourished status of one foreign group or another—the list goes on. Certainly the values we learned in the early years of our lives stay with us and are difficult to dislodge. And we certainly do learn values from our parents. They teach us either overtly and deliberately or subliminally through our observation of their actions. Maybe this is one mechanism through which "the sins of the fathers are visited on their offspring for three generations."

Another source of our paradigms is the first community we join outside the nuclear family, that is school. Our teachers are certainly presented to us in a context that identifies them as authority figures and role models. The values they act out with the group and their interaction with the group further extend and add to our paradigms, particularly in the area of our relationship to outside authority. This also extends the lifelong exercise in conflict resolution. Perhaps before we ventured outside the family circle the phrases "but Mom said" or "but Dad said" were heard, not as the question relates to some trivial passing matter, such as which jacket one should wear given the weather forecast, but rather as it relates to values and lifelong behavior. "But, Mom said that boys should never hit girls. Boys are stronger than girls and it is their responsibility to protect them and take care of them." This can usually be resolved within the family if Mom and Dad are on speaking terms. A few years later it may be "But, Mrs. Grinch said that we are all the products of evolution; specifically, we descended from monkeys." This may be more difficult to deal with, particularly if there is an examination to pass at the end of the semester. We also begin to develop our paradigms about our relationships with our peers. What works and what doesn't. What we can expect as the consequence of certain kinds of behavior. What gets us our way and what gets us a sock in the chops. It is fascinating to watch a group of children at play to see these early paradigms at work. Which subgroup screams at the drop of a hat and at the top of their lungs? Which shouts? Which seemingly instinctively slaps and which hits with a fist? Right or wrong, boys and girls behave differently (for the most part) because they have learned what works and what they can get away with and which kind of behavior is most likely to get them their way.

As our education continues, the size and shape of our educationally acquired paradigms evolve in more sophisticated ways. Our paradigms also become more rigid. (Does an automobile *really* have to have four wheels and an internal combustion engine?) The most charismatic and

persuasive teachers have the greatest influence. People often choose life-
time careers in emulation of a particular teacher or become interested in
particular subjects because someone made the otherwise dull, fascinating.
The values we learned in early life may be shaken by the influence of the
arguments of our later educators.

Certainly the influence of our communities is profound. If you were
raised in Amana, Iowa, and your barn burned down, your expectation
would probably be different than if you were raised in a big city tenement
and it burned. In one case your neighbors are apt to organize a barn
raising. In the other you might worry more about looters. Different paradigms.

As your horizons expand, you develop a more global outlook. We are
all experts at foreign policy. Considering the profound lack of information
that most of us possess in this area, where else could our firm ideas come
from but from our paradigms?

One way or the other, we all hold religious paradigms. We may possess
strong religious convictions, which color our behavior in all areas of our
lives. Or we may be flaming atheists, and this set of beliefs, or nonbeliefs if
you prefer, also colors our behavior. (If you are sure you only go around
once in life, you look at things a bit differently.)

It is very obvious, then, that perhaps the most influential forces that
forge our paradigms are the communities that we are members of. These
include our families, churches, schools, fraternities or sororities, political
parties, universities, labor unions, neighborhoods, professional circles, and
informal social circles.

Changing Paradigms

There are three ways in which we can change our paradigms. One is by
experience or the introduction of new information. One is by revelation or
insight, and the other is more deliberate.

Introduction of new information is primarily the turf of the educational
system. Hopefully, we attend with an open mind and expect to acquire
new information and different points of view. Once we are out of this
environment, the acquisition process becomes more difficult. After all, we
have been given all the answers; we are now in the business of looking for
problems they can be applied to. This suggests that certain of our paradigms are pretty much set in concrete. They serve as filters and are used to
test data to see if it is the "truth." "No, that doesn't make sense" means

that I have tested the data you just gave me and it doesn't fit my paradigms; therefore it is faulty data. Our paradigms are no longer open and malleable. Sometimes it is difficult to decide whose paradigms are most inflexible—the M.B.A. just out of the university who knows that the old fogies (which may include the previous graduating class) are obsolete or the senior citizens who are convinced that nothing of practical value or moral worth is taught in the universities anymore. (We who are more enlightened appreciate that there is a lot of truth in each position.)

But it is obvious that paradigms once formed are difficult to change, if they can be changed at all. Changing paradigms by the accretion of experience is at best a slow process. If "reality" is changing or expanding at a brisk rate, but paradigms are slow to change, it means that most of us are behind the reality curve and falling further behind daily. This is what leads to the problem of some old-timers becoming obsolete, not in the "real world" or twenty years behind the times, all expressions that relate to paradigm obsolescence.

The second way that paradigms can change is by way of "revelation" or sudden insight. This is the paradigm shift that many speak of. The classic example of this kind of paradigm shift is the profound religious experience in which an individual suddenly or over a relatively brief period of time adopts an entirely different model of reality, a model that transcends all others and affects every aspect of the individual's life. On a lesser scale is the individual who adopts a cause of one sort or another that seems to reorder priorities in his life and overwhelm other considerations. These changes come about with a bit of a lurch. In fact this is the predominant characteristic of this kind of change. A lurching, discontinuous change in outlook that is more revolutionary than evolutionary. This is not necessarily instantaneous revelation, but may very well be a lurching perception or realization of the significance of experience that has been accumulating unappreciated for some time. This is the kind of experience that executives sometimes have that causes them to come into the office one Monday morning with a whole new outlook on their business. Maybe they were influenced by a book read over the weekend, but whatever the trigger, they have experienced a paradigm shift that in many cases changes the most fundamental aspects of their management style, their corporate values, and their leadership personality. "Whatever got into old Fuzzdome? He's a new man!" Yes, he is, in a very real way. And if *you* and *I* are uncomfortable with the new Fuzzdome (after all we have spent years figuring the old goat out and now he goes and changes all the rules), you should talk to Fuzzdome! He doesn't quite know what to do with himself either. But there is one thing of which he is certain. Some things have to change. And if he isn't quite sure how to change them, he'd better figure it out, and quick!

The third kind of change is less dramatic. There are many individuals who don't know paradigms from parachutes but do know that one has to keep up with the times. This does not just mean the latest technological advances and management trends—one must stay open in more fundamental ways. Yes, this person should become computer literate even if the tool of choice when he was in school was the slide rule. And when intelligent people talk, our hypothetical person feels he should listen. He might learn something. This is the classically open-minded person, a rare specimen.

As for you and me, we would be well served if we simply understood the paradigm paradox and remained sensitive to it. Paradigms are good, even necessary. Having a model of reality is essential to sanity. Knowing that tomorrow will be pretty much like today in its important features is essential to our mental health. And that knowledge is a product of our paradigm about life. Sharing a commonly held paradigm about life is a good working definition of sanity. Someone who, for example, worships the brussels sprout as the supreme intelligence of the universe is crazy, that's all.

We should also know that since paradigms filter information or even render it invisible, we should be careful in examining new evidence. Our paradigms just might need adjustment. If the Swiss watchmaking guild had been sensitive to the paradigm paradox, they might have looked more closely at the quartz crystal movement and their world would have been different today.

The Paradigm Imperative

One could argue that a paradigm shift in the minds of management is an absolutely necessary prerequisite to any change of significance within the organization. Now this seems obvious on the surface. Yet, how many managements attempt to bring about radical change in the behavior of others without a concomitant change in their own values, priorities, and behaviors? This is almost a definition of the "Fad of the Month" phenomenon. A top manager reads a book or a few articles. A new idea or outlook is intriguing. Well, certainly it's fashionable. Everyone at the club is talking about it. You're not "with it" if your company doesn't practice it. There's a new vocabulary to be mastered. So they send someone to understand the "technique" and direct that it be implemented and then proceed to go about their business smug in the confidence that they are now up to date. These are the programs that wither and blow away. These are the charades that cause the organization to say, "Here he comes again. Lie low, this too will pass away."

Meaningful change begins in the gut of the leader. The leader must have a vision so compelling that it cannot be delegated. He is compelled to participate. He may delegate pieces of the action to others. He will push and haul and jab to get others involved, but he will be personally and intimately involved from the very beginning—not because someone has said that his participation is essential to success, but because the vision is so compelling that he cannot *not* be involved.

We tend to get the cart before the horse in this area. We observe a strong correlation between executive involvement and success as we look around us. Therefore, we insist on the need for executive involvement. If the program fails, it's because we didn't have such involvement or at least not in sufficient amount. The point is, we have confused cause and effect again (this happens all the time). In the successful programs executive involvement was an effect not a cause. The cause was a fire in the executive's belly that would not allow him *not* to participate. That same fire made him the relentless, obsessed, constant pain in the organization chart and that is what made the program succeed.

We in the quality field used to say that we needed management's support. We did. And when we got it, somehow it wasn't enough. More recently we've said we need management's participation. We do. And that's not enough either. Both these points of view imply that the mission of change is ours and management has a role to play in it, albeit not a terribly big one—sort of like a celebrity name for your golf tournament. You need the draw, someone to play in the Pro-Am, stand around at the cocktail parties, and be interviewed on television. But when the tournament goes down, the celebrity is expected to politely recede into the background. Not so here. This is management's show. They need *our* support and *our* participation, not the other way around, and this usually comes about as the result of a radical paradigm shift on the part of the executive.

An old expression of our culture is particularly appropriate. Without using the word, it implies a profound paradigm shift. Let's use it here. The boss has got to get religion!

CHAPTER 2

The Culture Trap

JOURNAL: *I've been looking around this corporation of ours. Not a long time, but long enough to know that I don't like what I see. This newfound sensitivity of mine may drive me nuts. I have found myself going into offices and manufacturing departments that I haven't been into in a long time. Even the places I go regularly don't look the same to me. Things don't look and sound and feel like I thought they did.*

I've asked questions. Different questions than I've asked before. People look at me and mumble. This place is dead on its feet. Everyone here is as uncoupled from reality as I have been. I ask if they think our company is at risk. They tell me we've been here for over seventy years, most of them as either number one or two in the industry. I ask them if they feel secure in their jobs. They tell me that we have excellent benefits. I ask about our foreign competition. They tell me they're "Johnny Come Latelies." What do those answers have to do with the questions? I snapped at one of our guys who answered me that way. I bit him for responding the same way I would have a short time ago. He probably got his answer out of one of my speeches. I really ought to make an opportunity to apologize to him. My reaction didn't seem to bother him too much. Instead of shaking in his boots, he just stood there with a silly look on his face. Was he used to being treated this way? Was he immune? I've got to change the culture of this place, wake it up, instill a sense of urgency. Culture change. Everyone seems to be talking about culture lately. It's such a trite expression.

I called a staff meeting. Did I give them an earful! Foreign competition, declining market share, deteriorating margins, consumer complaints, the absolute requirement to change, change big and change fast. And did they give me an earful in return; silence mostly. Just what I should have expected. At least they managed to stay awake, although there were too frequent surreptitious looks at their wristwatches. Then there were those who were telling me how grand everything really was. But just a short time ago I was saying the same thing. I would probably have accused anyone who was talking as I am now of disloyalty. Then there was my resident "time and patience" maven. That guy would have sat down in the middle of Pompeii and preached the virtues of time and patience. We've got damn little time and I've got absolutely no patience. Then there's the wimp! If he agrees with me one more time I'll fire him on the spot. Maybe I should. Every time he yeses me and I don't show signs of throwing up, I lose a little more of the staff's respect. A couple of looks were easy to read. "Someone got the boss a new book for Christmas." "Wonder how long this will last." One, just one, seemed to know what I was talking about. It was that new gal. Bright. Only been around here a few months. Her look said that I wasn't telling her anything she hadn't seen for herself. Maybe she was just a little surprised that I had seen it. Wondered why I hadn't seen it before, why I was seeing it now. Most of all, wondering if I had the brains and the guts to do something about it.

These are all bright people, even the wimp. They're hard-working, even if we do a lot of the wrong things. I remember when each of them came aboard. They've changed. Why? Why is that woman different? Is it because she has been around long enough to see how we work but not long enough to lose her perspective? Could she help me? Can anyone help me? Maybe I should charter a committee, give a few of these a special assignment. Maybe this isn't as hopeless as I thought. How do I go about changing the culture of this place?

Culture Defined

One of the difficulties that executives have when faced with the need to change the culture of the organization is that they don't really understand what culture is. They also make the mistake of thinking that it is theirs to change at will and they proceed accordingly.

The *American Heritage Dictionary* defines culture as "social and intellectual information" and as "the totality of socially transmitted behavior patterns, arts, beliefs, institutions and all other products of human work and thought characteristic of a community or population." Sound like paradigms, don't they? These ideas are true—by definition, so to speak. Accepting these definitions, what makes the boss think he can change the culture of the organization? More specifically, if it needs changing who must change it and how?

Is it fair and helpful to carry this definition one step further? Can we say that culture is "a community's response to its environment"?

A Community's Response to Its Environment

A community is defined as a group of people living in the same locality and under the same government and as a social group or class having common interests. This would seem to apply to the people within a company. They are a community.

The government they live under is certainly a big part of their environment. Companies are governments of a sort. The senior executive establishes a structure of government, a hierarchy of authority. He creates staff service organizations. He decrees policies and enforces them with laws. He creates a system for channeling the resources of the community into areas of need. Civil governments do this with subsidies, taxes, and tax incentives. Executives do it with organization, procedures, and a reward system, very similar actually. A universal truth among all governments was well stated by an American politician. "Those who go along, get along." This is certainly true in business.

Do the members of the industrial community really share common interests? As long as they are members of the community they certainly do. When times are good, all seem to share in the prosperity of the community. They enjoy economic benefits, job security, respect in the greater communities of which they are also a part—the industry, the business community, and the cities and towns in which they live. Similarly, when times are bad, all share in the down side. Belt tightening, demotion, loss of jobs, scorn within the industry, and even a little chill within the towns in which they live.

The Executive and the Community

Now we pose an important question. Is the executive a member of the community? Let's apply the criteria. Is he under the government? Not really. He *is* the government. The policies he has established for others apply to him only to the extent that he chooses to follow them. He is invariably sheltered from the same rules that he has established for the community to follow. He probably doesn't even have his home in the same locality. He tends to have his office in an attractive part of a cosmopolitan area. His home is probably in a different neighborhood from his head-quarters' subordinates. He is separated in terms of social as well as physical distance. He tends to have his factories in areas where land is cheap, labor is plentiful, taxes are modest, and transportation available for moving the products of the factory to market. When his duties require him to visit the factories, he may travel in a corporate jet and be met at the airport by the senior site manager who arrives in the closest thing to a limousine that the town has to offer. The executive is probably accompanied by an entourage whose job it is to see to it that he sees only what he wants to believe exists. The plant has been cleaned from stem to stern to a degree that clearly isn't justified for the unwashed masses or it wouldn't need special attention for the visiting executive. The smell of fresh paint is so universally present in the plants he visits that he may think all manufacturing plants smell like that all the time. When he says with characteristic humility, "I can tell a manufacturing operation just by sniffing the air," he really means it.

Then there is his personal economic situation. If he is a CEO or one of the executive inner circle of a North American corporation, he may well have a multiyear contract. He certainly isn't living from paycheck to paycheck, like many of the other members of the community. If the statistics are to be believed, the disparity between his income and that of the rest of the employee community is the greatest in the industrialized world. He may well possess a "golden parachute," which further insulates his personal fortune from that of the enterprise that he is responsible for leading. If leadership is lonely at this level, it is because the executive lives in splendid isolation and is anything but a member of the community whose culture he would change.

Social and Intellectual Information

To the extent that culture is a community's fund of social and intellectual information, the executive is also a stranger. He and the community of employees probably operate from totally different funds of information about the company. His is factual and intimate as it relates to the operating parameters, strategies, and tactics of the business, while the community is typically deprived of much of this information. As any community requires a minimum amount of information, if it is not provided the community will invent it. The means of communicating it is informal and the contents are rumor, speculation, legend, and folklore. On the other hand, the executive's vision of how it is to work here and the attitudes of the rank and file of the organization are probably even less realistic. In the final analysis, since we all behave in accordance with the information we possess, or think we possess, and the funds of information are radically different, then in a real sense the executive and the community are living in two radically different worlds, each group appearing unrealistic, insensitive, and generally inexplicable to the other.

Man the Barricades

Culture belongs to the community. To even understand it, let alone be a part of it, one must be a member of the community. The community insulates itself from the would-be outside leader in many ways, some subtle, some not so subtle. A labor organization is a very overt way in which a community moves to insulate itself. It also puts in place its own leadership network, system of beliefs, traditions, and physical responses. Management further augments this isolation. After all, they are part of a different community and they also wish to separate themselves. One way that they do it is with "perks." Even among the company elite who are furnished company cars, not all are allowed to have the same level of automobile. Managers probably park in a different lot from the "employees," and if the senior executive parks in that lot at all, it is in a location that is further differentiated from the centurions of the organization. The executives may well eat in a private dining room from the rank and file, and perhaps the senior executive has further differentiated himself by enjoying a private area either within or removed from the executive dining room.

Community Paradigms

The previous chapter explored the subject of paradigms. Individuals possess paradigms. So do communities. The definition of community provided earlier referred to socially transmitted behavior. Those are the community's paradigms. Bring together a group of people, and they will bring with them

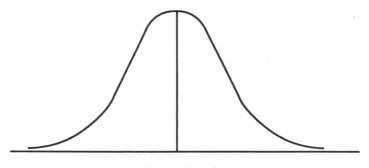

Figure 2.1. *Gaussian curve.*

their individual paradigms. While there is variety in these paradigms on a large number of subjects, the collection of paradigms is statistical in nature. That is, the distribution of individual paradigms has a shape, probably rather bell-like, it has a predictable degree of variation among its members, and it possesses a center of gravity of sorts; an average. Individuals contribute to the paradigm pool and they are also influenced by it. New members of the community quickly learn their version of reality by absorbing the paradigms of the community.

As asserted previously, a good street definition of sanity is a set of personal paradigms that conform reasonably well to those of the community. Certainly one whose paradigms don't conform is labeled an "oddball," at least. Oddballs tend to be ostracized. They make people uncomfortable. Consider the common statistical model, the Gaussian curve (also called the bell-shaped or the "normal" curve) (see Figure 2.1). The term "normal curve" is a curious example of paradigms at work. The suggestion is that distributions taking on other forms are abnormal and therefore not quite right—even though we know that many other shapes of distributions exist, and while they are not quite so common, they are certainly not "abnormal."

The center of the distribution represents the norm for the community (there's that word again). The spread represents the variation of the individuals about the norm. We expect variation. But we also sense limits on the variation that we expect. We might think of this expected variation as respectable variety. In some situations the variety about the norm has no value implications. Being near the right limit is no more or no less desirable than being near the left edge. In other matters there are value implications. The right side of the distribution might represent superiority in the eyes of the community, the left side inferiority. Where a value scale is present,

Figure 2.2. *Community of communities.*

most of us would prefer to be displaced toward the valued end of the scale. But whether the scale has value implications or not, we all wish to be located *within* the range of normal variation. To get outside the accepted range of paradigm variation is uncomfortable and perhaps even dangerous.

Subcommunities

This is why communities at rest—that is, without some agitating influence—soon begin to sort themselves out into subcommunities. Various ethnic groups begin to withdraw into physical communities. Surely there are some physical reasons that this occurs, but they are no more powerful than the social reasons. These subcommunities sort themselves out along lines of political parties, union attitudes, style of dress, accents, attitudes toward one another, religious beliefs, a whole host of social parameters. They create subcommunities that reduce the variation among individuals. Soon these subcommunities become the real communities to which the members belong and the parent community becomes an abstraction.

The central tendency of these subcommunities is probably not radically different from that of the collective community. Even within prison populations there are limits on the accepted range of criminality that is tolerated. Prison is an unusually dangerous place for perpetrators of certain types of crimes. "I'm a murderer, but I'm not a _____" is a common attitude. What is different is that within each subcommunity the variation about the norm is dramatically reduced. It was variety in the general population that caused these groups to coalesce into subcommunities in the first place. (See Figure 2.2.)

People tend to select for association others whose paradigms they share. People making the hiring decisions within a company also tend to hire people of backgrounds, tendencies, and values similar to their own. People approve of people who think as they do; they even value them. "Outliers," people who are found to be outside the limits of the population, have a difficult time. At best, they are the "loners" within the community. Perhaps they can't even find gainful employment. "Not a team player"—while sometimes a judgment that can be taken literally—often refers to one who lives in relative isolation from the other members of the community. When management wants the community to shift its paradigm, is it any wonder that no member wants to be the first to move? They certainly are reluctant to move outside the limits of respectable variety even if intellectually they see the logic of the appeal. This is the ultimate peer pressure. Should we be surprised, therefore, in the industrial setting if the engineering department has paradigms that vary significantly from those of manufacturing, purchasing, and sales?

Culture belongs to the community. No one outside the community can change it. But the right actions outside the community can help it to change. If culture is a community's response to its environment, then there is hope. The executives who live outside the community are nonetheless largely responsible for the environment in which the community lives and to which it responds. Change the environment and do it adroitly and the community may well decide to change its culture. How much and at what rate depend on many factors.

Changing the Culture

Organizations comprised of many subcommunities, and that probably includes most large organizations, are in a sense schizophrenic. Their corporate personality has disintegrated into a large number of subpersonalities that either take turns dominating the corporate personality or assert themselves in functional activities or geographic areas. Changing the corporate culture in a deliberate way requires several types of action.

First, the executive must get in touch with the communities and their paradigms. Note the use of plurals here. If the executive looks for the corporate personality and its paradigms, then he is probably wasting his time. There probably isn't one. The executive is not required to accept these paradigms, only to understand what they are and be empathetic with the people who hold them. They exist for a reason. They are that community's response to its environment. In that sense they are logical and rational.

The executive must lead in the reintegration of the corporate personality. He must create an environment in which it is natural for people to subordinate the interests and perspectives of the subcommunity to those of the parent community. He cannot require that they make this shift. He *can* require that they rearrange themselves in ways that create different subcommunities that are more relevant to the corporate mission and therefore weaken the bonds of the former structure.

Finally, the executive must provide the communities with information. The bias should be away from providing that information which is essential to doing their jobs and toward providing all the information that it is permissible to share. Information is what creates the bonding within one community or the other.

Getting in Touch

How can the executive change the culture if he is out of contact with it? He may think that he understands it, but after all, he is observing it remotely and through his own paradigms, which in their own way are no more realistic than those of the community. In many ways they are probably less so. Getting in touch includes those in the offices as well as those in the factories. Getting in touch is difficult and in many ways dangerous. If the executive's actions seem phony, he will have reinforced the community's resistance to penetration and made matters worse. The executive must proceed very carefully in this area.

Parking in the employee lot one day and sweeping into the factory with his necktie off and calling everyone by his first name (having spent the night before memorizing all the names) is probably not the best thing to do even though the individual elements of that plan might be great in their time.

Skip Level Meetings

The executive might initiate a series of small-group meetings with representative members of the significant subcommunities. To avoid throwing the community into "culture shock," he might proceed from his inner circle out. He should meet with them on their own turf and under their working conditions, not his. He must absolutely avoid the artificial, and that means the community's view of what is artificial, not the executive's.

Many people will be stacking the deck. As he's a bit out of touch to begin with, he is probably in a disadvantaged position to detect deck stacking unless he is very deliberate in his sensitivities. He might start out by making it incandescently clear to those who are making the arrangements (selecting the attendees, meeting location, etc.) that stacking the deck will be considered an act of insubordination and will be dealt with accordingly. After all, he is still the boss and it pays to reiterate that fact occasionally when he's working the softer side of the equation.

He is going to get no more candor than he gives. If he had all the answers, he wouldn't be having the meetings. Let that fact show. He's not telling at this point, he's listening. He has some real concerns, or he wouldn't be going through this procedure. He should start the meetings by expressing those concerns. There are things in the community that he doesn't understand. He should express these concerns and ask for input. It is important that he ask the communities to provide information on the issues where they are expert and he is ignorant. He is concerned about organizational behavior. He should define the behavior that bothers him as best he can and in terms that they can relate to and then ask a simple "Why? Then he should do three things: listen, listen, listen. He should play back to the employees what he's heard to make sure he's heard it right. He should be sensitive to the cultural boundaries of the group to which he is talking, or rather listening. Preferably he should be the only outsider in the room. If the plant manager is not a member of that community, he shouldn't be there even if it is his plant. At this moment it's their plant. No recording devices! He can take his own notes. He must not argue with the comments he gets and he cannot defend the status quo. He is there to understand their point of view, not sell his. He should make only three promises and be scrupulous in standing by them. He should promise to listen. He should promise to be nonjudmental. He should promise that nothing evil will befall anyone because of what is said in these meetings, and that includes retaliation against anyone whom individuals feel like complaining about. This is not a witch hunt. He should not promise anything beyond this. He is in no position to, and the employees know it. Another casual promise will not reinforce his credibility.

Facility Tours

Plant tours are a good thing to do. The executive should consider making a plant tour following his skip level meetings—*after* not *before*. That way he might recognize people and exchange a few nods, but no claps on the back or first name greetings. The executive and the employees are not lifelong buddies, and both know it. Noblese oblige is toughest on the dignity of the peasant. The entourage should be kept to a minimum. The plant manager should be included. After all he *is* the plant manager and his position must be reinforced. By this time he is justified in feeling a bit apprehensive and the executive should use this opportunity to reinforce the plant manager's status in both the minds of the workforce and that of the plant manager.

The executive might include one of his trusted staff, a politically neutral person, in the touring group if only to keep the plant manager busy at strategically important times so that the executive can exchange a few words with the employees in an unencumbered way. Not that he's spying on the plant manager. That notion cannot be allowed. He might reach out with a bit of the unexpected, like asking a line operator if the factory has been cleaned up just for his visit. The shock of being asked such a "real world" question will probably get him an honest answer and maybe a bit of new-found respect. It will also signal that he knows the difference and might not be as out of touch as the people think. If factory priming is common (it probably is), he can later make a general policy statement expressing his appreciation for their consideration of his delicate sensibilities but gently threatening the life of any plant manager who in the future stacks the deck in that particular way. He might even suggest that the money currently being spent in that manner be redirected to housekeeping matters of more benefit to those who work there day in and day out.

Why Bother?

There are many ways that a perceptive executive can use these occasions to "reach out and touch someone." His principal motivation here is to reach into the community and make contact. He's still the boss. An overly casual approach will only reinforce the gap, not dispel it. But he does possess an umbilical scar just like the rest of us and it doesn't hurt to acknowledge it. And there should be no souvenir photographs or any of the other not so subtle tactics that remind all they are in the presence of royalty—they aren't!

These are just some of the opportunities the executive can create in order to get in touch with the organization. But there is an important point to be made here. Normally we think it is the organization's responsibility to stay in touch with the executive. The executive directs, the organization responds. Is this then an appeal to "touchy feely" management? Not at all.

In the role of change agent the executive is not performing as a manager, but rather as a leader. And the leader must get in intimate touch with those whom he would lead.

Too often executives are insensitive to the important differences in these two roles. Therefore, they visit plants and operations of one kind or other and proceed to stand behind lecterns and tell and explain and direct. But they are telling about these "realities" from the point of view of their paradigms, which are not those of the community they are addressing. They explain the facts as they have been filtered and interpreted by their models of reality. The community's paradigms may well make the executive's data invisible to them and that might explain the blank looks and nonresponsiveness of the audience. In the executive's directions he is trying to move the community toward objectives that may be invisible or irrelevant to the members.

The first responsibility of the would-be leader is to get in touch with those whom he would lead. He cannot expect, let alone demand, that they come to where he is. They probably don't know where he is coming from, let alone where he is. They probably don't know how to get where he is, even if they want to be there, and finally they may well have no interest whatsoever in being where he is. If the executive would lead, then the executive must go to where they are, understand their perspectives and paradigms, and learn what makes them tick (a particularly useful metaphor here). Then and only then, having merged with the community, at least to the extent of joining them where they are, can the executive begin to influence them in the directions that he would have them go. This is *not* "touchy feely" management. This is a necessary element of leadership.

Reintegrating the Corporate Personality

An organization at rest tends quite naturally to sort itself out into subcommunities. In this sense it is like some liquids left quiet overnight that aren't quite emulsified. The executive is interested in weakening the bonding of the members to the subcommunity so that they can be bonded more strongly to the parent community. This is necessary if a corporate culture is to be reinstated based on common interest. Common interests must be established among the members of the parent community that are stronger than those of the subcommunity. The executive is not trying to destroy the subcommunity, or he shouldn't be anyway. As in any community, ethnic diversity is to be encouraged and is a source of enrichment to all the members, but only if these ethnic groups share a healthy set of common interests that transcend local issues. The best way that the executive can accomplish this is to agitate the community, quite literally shake it up. He must seize on opportunities to bring people together in settings that relate them to the larger community and, if need be, subtly discourage gatherings that encourage insularity.

Company picnics and holiday events can be substituted for department events or at least be included in the overall program. These also provide opportunities for the executive to mix with the organization in an appropriately informal way. The company can support Corporate Challenge and community service activities and encourage participation of members of the corporate community. A favorable image for the company in the communities in which it is located reflects on all employees of the company, not just one segment. Company-sponsored celebrations are a particularly effective way of strengthening this dimension of the company. Several corporations regularly gather people to share, but perhaps more important to celebrate, accomplishments by teams of employees. As these teams are, in many cases, interfunctional, this further breaks down the insularity of the subcommunities within the corporation. Any activity that creates or reinforces the identification of the individual employee to the corporation contributes to the integration of the corporate personality.

Organizational devices assist in reasserting the corporate culture. Matrix organizations can draw the members of the organization into broader associations than strictly functional organizations. If functional structures have prevailed for a long time and have become the basis for overly vertical orientation, then product-based organizations can help break down functional barriers, particularly when the members of the product organization are physically moved together.

Ad hoc teams to attack problems or take advantage of particular opportunities have many advantages in their own right. These team approaches also have the advantage of bringing together interfunctional teams in intimate situations that can help break down the barriers between functional organizations. Successful team activities create very strong bonds between the members of the team. Team celebrations and recognition enforce these bonds.

CHAPTER 3

The Work of the Leader

JOURNAL: Some cynical individual said that leadership is the art of finding out where the people are going and getting out in front. Maybe that person wasn't a cynic at all. Perhaps he just understood the process. I'm coming to realize how out of touch with the organization I am. I don't think that's usually a big problem. In this age of specialization I have my job, others have theirs. I'm pretty good at doing the C.E.O. things, the management end of the business. But moving this company onto a different path requires something else, leadership I guess. But are management skills and leadership skills really different? If so, how? The people who report directly to me don't seem prepared to provide what's needed either. They've been selected for their managerial skills and they're good. But doesn't management include leadership? If I want to change the direction of the organization, then I'm going to have to position myself to provide leadership, whatever I decide that is. As that fellow suggested, somehow I will have to get out in front where they can see me and follow in the directions in which I move. I've been standing on a hill waving my arms and shouting, "Over here." They can't see me for the dust and can't hear me for the noise of the battle. Besides, at the moment their first interest is in ducking the next arrow to come their way. Not a bad analogy. How do I get them out of the moment-to-moment survival mode and onto the offensive? How do I change the rules of the engagement? I guess my first challenge after all the intellectualizing is to figure out how to get out in front. When I get there, I had better have figured out the difference between management and leadership, if there is any.

The Management Challenge

Management is the process of allocating scarce resources to the achievement of enterprise objectives. Therefore, managers have two outcomes to focus on. First is achievement of the objectives of the organization. Perhaps the managers participated in the formulation of these objectives, perhaps they didn't. That in no way changes their responsibilities. In this era of participative management, it is sometimes felt that if an incumbent at any level did not participate in the formulation of the objectives, then he is not quite so attached to them. Psychologically this may be true. But the notion follows, in a subliminal way, that as a result the managers don't have quite the same obligation toward their execution. Somehow those objectives are not completely legitimate. Nothing could be further from the truth. Managers manage. They execute. If they find the directions repugnant to them for some reason, then they must conclude that they did a bad job of selecting their boss and dissociate themselves from the enterprise. There is no honorable alternative. And rumors to the contrary notwithstanding, honor is an important attribute of a manager.

The second responsibility of a manager is to achieve the objectives efficiently, that is, with maximum economy. This is the "allocation of scarce resources part." The manager has been given stewardship over a bundle of scarce resources. Depending on the level of management at which he finds himself, that bundle may include a number of people; some machines of various types, ranging from word processors to whole factories; some raw materials or operating supplies; some floor space—the usual "M's"—men, machines, money, materials, methods. All of these are available in finite amount, usually penuriously finite to hear most managers tell it. The manager's task then becomes simply, or not so simply, to allocate these limited resources to the achievement of the objectives that they have been charged with accomplishing.

There is one more scarce resource that often gets overlooked. Perhaps it's because the word doesn't start with "M" and therefore ruins the alliteration. That most vexing of all scarce resources is time. And the time resource has some interesting characteristics.

Consider this analogy. The manager's task is like constantly managing a pan balance. In one pan is the objective that the manager is responsible for accomplishing. In the other is placed the necessary amount of resource to balance that task. Time can be considered the fulcrum point. If the fulcrum point, time, is in close, then more resource will have to be placed in the pan to balance the weight of the task. But what manager ever had only one objective to attain? The manager has many pan balances to manage. He also has one stockpile of each scarce resource to be allocated. We are now developing a model for the dynamics of management. But we're not through yet.

Not all resources have the same degree of scarcity. Some are more valuable than others. Some are more weighty. Some can be traded for a quantity of another, some can't. To a degree management can buy more time (labor hours) in the short term with operating capital in the form of overtime (at a scale factor of time and a half or double time for overtime compensation). Fixed capital investment can often be substituted for labor, but it takes time to make the substitution. You want to "buy time"? (An interesting expression under the circumstances.) You can often do so through the lavish insertion of resources. But although this may seem wasteful, perhaps it's not. Presumably time is the more scarce resource in this situation.

Given the tradeoff notion, we can modify our definition of management if we wish, as follows: The process of allocating scarce resources to the achievement of enterprise objectives with the minimum depletion of the total weighted resource. It's beginning to appear that management is a full-time job!

Now someone comes along complaining about some other issue for which the manager has neither a pan balance nor a resource. Perhaps one should not be surprised to get in response from the harried manager a somewhat dazed expression and a nod or two, and then the manager runs off to play with the pan balances before they get all out of kilter. Maybe it's also a bit like the juggler balancing spinning plates on sticks. He's got all of the plates that he can keep spinning, thank you, and you want him to stop and listen to a dissertation on "Corinthian Columns: Their Various Heights and Descriptions." The next sound that you hear . . .

There have been many colorful descriptions of the practice of management. One of the more printable is: To have been trained in the organization and conduct of cattle drives only to be dropped into the middle of a stampede.

The Leadership Challenge

In his address to the American Society for Quality Control's (ASQC) 1986 Annual Quality Congress, leadership was defined by Bill Ouchi as "the ability to cause others to follow into areas of uncertainty." This remains a very elegant thought.

As hectic as the management task is, it is focused on the "here and now." It deals with resources with which the manager is familiar, risks that are usually less than life threatening, and rewards that one can envision as realistic. Management is a practical art that deals with the familiar. In its purest form it is the efficient pursuit of ordinary and accepted goals using ordinary and accepted processes. It imposes little demand on the leadership ability of the manager. That is not to trivialize the role of the manager. The manager has a big bone to chew on.

The authority that the manager draws on in fulfilling his function is statutory authority, the authority of position. It is conferred by a third party, implemented by rules, regulations, and procedures, and enforced by the power to reward and levy sanctions. Its objective is conformance.

Leadership, on the other hand, has as its objective causing others to accept the risks of venturing into the unknown. Demonstrating a remarkable grasp of the obvious, one might conclude that the most frightening thing about the unknown is that we don't know what's out there. But no soldier ever charged a machine gun emplacement because it was in his annual objectives. Ordinary people don't risk their lives because they are superbly managed. They do it because they're superbly led. Pursuing the military example, maybe that's why platoons have leaders, while companies and batallions and regiments have commanders. The lowest-ranking officer in the army has the task of causing others to follow into areas of uncertainty. Considering the mission of any army, it is appropriate that the newly commissioned officer must first demonstrate his ability to lead before he is given the opportunity to manage. Note also that it was management that brought that platoon to its place before the enemy. It was management that assembled them, trained them, equipped them, transported them, and supplied them. Without management there is no army. But without leadership there is no victory.

Characteristics of Leadership

What are the characteristics of leadership? In a speech to the Veer Foundation in the Netherlands in 1982, Joseph Jaworski, Chairman and CEO of the American Leadership Forum, cited five elements that he thought were key to excellent leadership.

First, the successful leader must have a compelling vision. Excellent leaders are concerned with the organization's basic purpose: why it exists; what it should achieve. They are not preoccupied with the "how to," or nuts-and-bolts, aspects of the operation. They see themselves as leaders, not as managers. The driving vision has an almost spiritual quality to it.

Next, the successful leader must be powerful, not through dominion or control or manipulation, but through the capacity to mobilize people and resources to get things done. Rosabeth Moss Kanter, a leading sociologist, describes the powerful leader not as an autocrat, but rather as a sensitive, empathetic, compassionate, empowering person—not so much an exerciser of power as a provider of power, a dynamo.

The third element cited by Jaworski was that the leader must exemplify the highest values of the organization. To earn trust the leader has to be authentic. He must come across like a good musical score. The words and the music match.

Next, the leader must provide breadth and risk-taking entrepreneurial imagination for the organization. He must see things in a fresh and different context. He must be able to "recontextualize" a situation.

Finally, in Jaworski's view, the effective leader is a transforming leader. Transforming leaders are capable of directing people through fundamental change—personal, institutional, and societal. The leader "engages with [his] followers, brings them to heightened consciousness, and in this process converts many followers into leaders in their own right."

For an understanding of leadership one could do worse than refer to what is perhaps the most thought-provoking book on organization and management ever written by a practicing executive: *The Functions of the Executive* by Chester I. Barnard. It may be more widely read today than upon its initial publication almost fifty years ago. In it Barnard defines leadership as "the power of individuals to inspire cooperative personal decision by creating faith." We don't hear much today about creating faith. Perhaps that's why schools don't get high marks (pardon the pun again) for creating leaders. In fact, we probably find the idea mildly embarrassing. After all, management is about "firmer stuff." Barnard goes on, "Faith in common understanding, faith in the probability of success, faith in the ultimate satisfaction of personal motives, faith in the integrity of objective authority and faith in the superiority of common purpose." Leaders create faith.

Barnard speaks of morals, not meaning any particular set of ethical values. Rather, he refers to the forces within individuals that tend to control their behavior to keep it consistent with their personal values. In other words, morals serve as personal flywheels that resist the derailment of the individual by the bombardment of short-term forces. When this tendency is strong and stable, there exists a condition or responsibility. Responsibility, as Barnard uses the word, is the power of a particular private code to control the conduct of the individual in the presence of strong contradictory impulses. Responsibility is the strength of the moral set. Responsibility, then, equals steadfastness, Deming's constancy of purpose. Barnard asserts that all of us possess several, if not many, moral codes—religious, patriotic, familial, business, etc. Persons differ therefore not only as to the quality, complexity, and relative importance of their moral codes, but as to the strength of their sense of responsibility toward them.

Leaders have the faculty and the responsibility of creating moral codes for the organization, which serve as value structures, organizational flywheels. The author finds it interesting that the words "moral" and "morale" have a common root. Organizations with strong, coherent value systems have excellent morale.

Barnard asserts that the important distinctions of rank lie in the fact that the higher the grade, the more complex the moralities involved and hence the need for skill in resolving the inevitable conflicts. When various sets of values come into conflict, and they inevitably will, the manager chooses between them. He places one above the other, thereby strengthening one value and depreciating the other. The leader, however, invokes creativeness. That is, he creates a *moral* basis for the solution of moral conflicts. He either substitutes a different action which avoids the conflict or provides a moral justification for exception or compromise. This creative function as a whole is the essence of leadership. The structure of organizational values is the spirit that overcomes the centrifugal forces of individual interests. It infuses the subjective aspect of countless decisions with consistency in a changing environment. In other words, the leader resolves value conflicts by eliminating them if possible, but if not possible, the leader creates overarching values so that conflicts are resolved instead of one value being subordinated to the other. Resolution is a win-win situation between opposing values; choosing between is win-lose.

Ouchi, in his remarks to the ASQC congress in 1986, referred to Barnard's work and, paraphrasing only slightly, observed that leadership in all cases was based on the personal integrity of the leader. He observed that integrity didn't imply any particular set of values, but rather a coherence of personality, the state of being "together." A person who is together (integrated) has a kind of transparency of personality. He always shows the same face to all people in all situations. This quality is an essential characteristic of effective leadership.

One of the several definitions of integrity, the one to which Ouchi referred, is "the quality of being whole or undivided; completeness." It is also the logical outcome of a process of integration, "to make whole by bringing all parts together." It follows that an executive cannot provide coherent leadership to the organization until his set of values (or morals, as Barnard would say) has been integrated with his other moral sets. This is done by creating an overarching algorithm which reconciles inevitable conflict to the satisfaction of both value systems. In the presence of this coherent value system ambiguity disappears. People do not have to choose which of the conflicting values are to be served. They have been integrated and can be served as a whole.

Management or Leadership?

Managerial ability and leadership ability are not mutually exclusive. They exist independent of one another. An individual can be high on either scale or both or neither. At the same time, most positions in the hierarchy of any organization require a bit of each, or at least would benefit from a bit of each. If a manager is put in place over a group of people, it follows that a modicum of leadership skill would assist in moving the work of the group forward and probably make the trip more satisfying and fulfilling for all concerned. On the other hand, most group tasks can be executed in a satisfactory way with modest amounts of leadership. Management skill alone will usually suffice. Managing a team of office workers in carrying out routine and repetitive tasks probably does not call for an inordinate amount of leadership ability. On the other hand, the person heading up a fire-fighting unit or a Coast Guard rescue vessel should probably be high on the leadership scale.

It would seem that overseeing the doing of unchanging, routine, safe, generally accepted work in a benign environment does not place many leadership demands on the person in charge. However, this kind of work requires more than the usual amount of managerial skill. Processes must be honed for maximum efficiency, people must be supervised to assure their attendance and diligence to dull duty, quality of output must be managed. In short, efficiency is the managerial product in these settings and that's a big challenge. All group endeavor will benefit from managerial as well as leadership skill. However, some kinds of work demand one over the other. Occasionally work comes along that demands a great deal of each. Thank goodness, not often.

JOURNAL: Moral authority! That's the basis of excellent leadership! Moral authority! Managers exercise statutory authority—rules, regulations, practices, deportment. They possess the means to enforce this authority through rewards and sanctions of one kind or another. But the basis of leadership is moral authority. People follow leaders because they choose to, not because they are required to. They can always "dog it" if they want. The result of good leadership is stretch—more, better, courageous, diligent. You don't get that by exercising rank. You get it because people choose to join up. If they believe that you're asking more from them than you are from yourself, they won't give it. If they sense that you are asking them to give what only they have to give and that you are giving what you have to give, they might. If they sense exploitation or hidden agendas, they won't. If they believe that what you're showing them is what they will get, they might. If they sense that you are pawning off the dirty work on someone else, they won't. If they see you doing your own dirty work, they might. If they sense that you are trying to motivate them to put their necks on the line so that yours won't be, they won't. If they see your neck on the line, they might choose to join you. Do some of these things and they might. Do all of these things and they will.

I remembered 1951. Korea, that was my war. Read in the paper about a small Turkish unit in the U.N. forces. Surrounded, cut off, out of ammunition, outnumbered seven to one. What did they do? They fixed bayonets and charged! Fought their way out. WE CAN DO THAT TOO!

A Summarizing Thought

It may be fair, although perhaps an oversimplification, to think of management as the process of using statutory authority to administer the allocation of scarce resources to the accomplishment of enterprise objectives. Statutory authority involves position, laws, regulations, procedures, together with the power to reward and penalize.

Leadership, on the other hand, is the process of exercising moral authority to cause people to strive beyond their normal limits. That "striving beyond" may take the form of risk of personal injury or failure. It may mean extraordinary effort in terms of hours or exertion. It may require the investment of one's ego and self-image in subordinating self-interest to a higher and nobler goal. None of this is available to the manager. Statutes and procedures can't command it. Rewards of the type that the manager can offer can't bring it forth. Threat of penalty can't extract it. Great leaders of history have never offered much in the way of rewards; at least not in managerial coin. Rather they have offered that ". . . death and sorrow will be the companions of our journey, hardship our garment; . . ." (Churchill), and ". . . pick up your cross and follow me . . ." (Christ).

Moral authority flows from the leaders' compelling vision, their power to mobilize, their living example, their willingness to put their name and their place in corporate history on the line. "Blood and guts" means their blood and their guts, not just yours—in short, the ability to create faith in the collective enterprise.

PART II

Implementing the Vision

CHAPTER 4

Vision—Or Just Seeing Things?

JOURNAL: I've been agitating for some time now. In every meeting, at every opportunity, I manage to bring up my concern for the future of the company. I feel like a street evangelist. But a little evangelism may be what we need. The response continues to be underwhelming. Some people are beginning to show a little irritation. Maybe that's good, I told myself. Finally I had a "heart-to-heart" with one of the staffers while we were together on the company plane. Well, not exactly. He talked and I listened. Hasn't been around too long, four or five years. The way he came on I didn't expect to want him around much longer. Bright and independent as a hog on ice. Not impudent, it's just that I don't think you should ask his opinion unless you really want it. Young enough to get a job somewhere else, I guess. A good observer. Wish I'd had his input before. Lots of good ideas in this organization if you can get at them.

Talking out loud, mostly to myself, I asked what I had to do to get this organization moving. I sensed he'd been waiting for this opportunity. Like a shot he came back and asked, "What do you want us to do?" I sensed a lot of feeling under that question. He caught me off guard. After a bit of dead air (I couldn't think of what to say), he went on . . . quite a speech.

41

"We know you're concerned about the company and its future. We know that you are convinced we must change. The people here really do care. After all, you've got people with thirty, forty years invested in this company. Don't tell them about the financial community and the shareholders. The biggest shareholders you have are the employees of the company. They not only own stock that they aren't allowed to sell like the other stockholders can, they've also got their past and their future invested. They haven't any stock options or golden parachutes. All they've got is a pension plan that they've been counting on for more years than the younger among us have been alive. They believe you. They're scared too, whether they admit it aloud or not.

"But what do you want them to do? Frankly, all of the angst without action serves mainly to convince them that you don't know what to do. They feel like ants floating down a river on a log. What business are we in? Where do you want to be in five years? Surely we've got five years. You've got a lot of great soldiers out there. Frankly, they need a flag out front to guide them. They're as good a group of people as there is on planet Earth. They're not the problem. What do you want them to do?"

That really ticked me off! Started to tell him so, then bit my tongue. What he was saying without using the words was that it's leadership they need, not exhortation. There's that word again. Maybe that's what their silence has been all about. So I challenged him a little (after all, I'm still the boss) "Okay. What do you think they need first or most?" I admit I was daring him at that point. One snotty remark and he's history! He fooled me. His answer—

"A vision." That's all? "No sir" (now he gets polite) "but you asked what they need first or most. I really believe what they, I should say we, need first and most is a vision. A common vision. We do have a lot in common. For one thing we're all employees. Maybe you and the vice-presidents should stop referring to 'our employees.' You don't own us. I think what you mean is 'our fellow employees'—we're in this together. That's the biggest thing we have in common. We're proud of this company and what it has accomplished. We're all convinced that the company still has a lot of greatness in it. We like 'Rocky' stories. Participating in a world-class comeback would be the greatest high in the world. But first we need a vision." I've read a lot about vision statements lately. Apparently he has too. I graced him with a smile and a moment of eye contact. "Good point," I said, "maybe I should charter a committee of executives to develop a vision statement. I would even be willing to chair it" (magnanimity on display). He did it to me again; thought he was going to lose it entirely.

"Please, no! We need a vision, not a vision statement. I guess statements are needed to express the vision at some point. But the vision must precede the statement. The statement doesn't make the vision happen. The vision makes the statement possible, and credible. And visions are formed and given life by leaders, not by committees. Sir, what is your vision for this company? Talk to people. Pick their brains. Find out how they feel about where we are and where we should be going. Try out your ideas on people to see if they pass the test, but in the final analysis you must be the visionary. Managers execute ongoing programs. We've got managers. But we need change. Major league change. Change requires leadership first, then management. We need a vision that's big enough for the followers to sign on to. One that's worthy enough for them to stick their necks out for." Asked him where he'd learned so damned much about leadership. The first time that he'd mumbled that day. Later I asked my human resources VP for a background on the guy. She said that among other things he had been decorated in Vietnam in some kind of action or other. Yeah, that was a good place to learn about leadership.

The Corporate Vision

There are several elements to a corporate vision. There is the mission, the strategy, the body of principles and values, and the organizational priorities. The mission defines why the organization exists, it's reason for being. Strategy has to do with planning the conduct of a large-scale campaign. In this case it relates to how the organization intends to carry out its mission. The values define the higher body of ethics and principles with which the campaign will be carried out. Priorities set the tactical rules. These elements are always present whether formally defined or not.

People can always tell you why the organization exists and who it is to serve. It's just that they might not all tell you the same thing. Lack of definition here invites incoherence into the conduct of the business. People with different ideas of the mission make different decisions along the way. There is always some sort of strategy. It may be good or bad, loose or disciplined, but there is some guiding idea behind the conduct of the operations. If the strategy is not defined, people will assume one or define one of their own, but there will be one. The difficulty in defining strategy from the foxhole, however, is the lack of visibility. Limited vision gives rise to poor strategy. By values we mean the standards of the organization. There are always standards even if they are to the effect that there are no standards; anything goes. There are always priorities but they will be situationally driven if not defined in advance.

The difference between an operative vision and just a pipe dream is a commitment. That commitment is given voice in the definition of the mission. A mission exists when the operation is one place, so to speak, and determines to go somewhere else. If a company is positioned exactly where it wishes to be and has no intention of being anywhere else or doing anything different, then we could say that it has no vision. One might question whether such a company can survive without a vision, but there's little doubt that it can at least remain an also-ran if that fits its reason for being. The basic point is that one doesn't require much vision to look down and see the tops of one's shoes. Neither does a company. But seeing over a far hill into a land where you've never been requires a great deal of vision.

Mission, strategy, values, and priorities are sometimes difficult to separate from each other. This need not be a big problem. Many mission statements are clearly visionary and often embody priorities and values that seem a lot like strategic objectives.

The Mission

The mission defines the organization's right to exist. It's the ultimate standard for judging its long-term performance. At the end of the road the ultimate test is whether or not you accomplished your mission. But corporations are created to exist in perpetuity. There is no end to the road. In most cases, for a corporation to reach the end of its road is the ultimate failure. It still has a mission. In this case a mission without end. The mission statement may be grand and inspiring or understated to the point of being spartan. Either can be appropriate. Much depends on the style of management and the nature of the company. But in any case, the mission must strongly reflect the vision. As a minimum, there are three essential elements to a corporate mission:

> Product
> Served market
> Positioning

Product

What is the product of the organization? Is it a physical product or a service?

In defining the product, entertain the idea for a moment that all that is ever provided by a corporation is a service. The physical product is merely the package that carries the service to the ultimate consumer. When one buys a can opener, consider that one is not buying it out of pride of ownership: "Look, friends and neighbors, I own a *can opener!*" One is buying it mainly because he or she regularly buys foodstuffs packaged in cans and needs a convenient way of getting inside. The former method of accomplishing this task dulled the hatchet and did grievous damage to the product. (Maybe that's where creamed corn came from.) Now since the consumer is justified in assuming that your product will provide the promised service, he may become interested in some other aspects of the physical product, its cost, its attractiveness, its portability, etc. Since you, the manufacturer, cannot differentiate yourself from your competitors by merely opening cans, all of you can do that, you may choose to compete on an additional basis. But while the service is presumed, it is not to be treated casually by the manufacturer. The history books are full of companies that took their service for granted and allowed it to depreciate while attending to secondary considerations.

Consider that some marketing genius suggests that you scrimp on the quality of the materials and processes used in creating the more mundane aspects of your can opener, so that you can afford to apply an exotic decoration that has demonstrated sex appeal with the domestic engineers in our focus groups. The marketer should be drawn and quartered. The tactic will work for a while. After all the customers *presume* that it will open cans. Over time, the whole can opener industry begins to compete on the basis of styling. Then a new competitor comes along. This one says, "See this can opener of ours? Ugly isn't it? Well, it's not exactly ugly, but it's definitely not modern art. We prefer to think of it as 'high tech.' But what it lacks in beauty it makes up for in heavy. That's because it uses the same motor that is used to lift concrete to the tops of skyscrapers under construction. It'll take the top off an oil drum! And no aluminum foil blade either. Indestructible. How did we do it? Revolutionary concept! We saved the money on design and fru-fru and put it on the inside." What happens? All of a sudden ugly becomes fashionable and tailfins become a national joke. Some companies went into business to make buggy whips. Maybe they still do. Others were in business to design and manufacture speed control devices. They now supply the automotive industry.

A mission statement must define the product or service that the company intends to supply. This establishes boundaries for the business in one dimension of its activities.

Served Market

This establishes the boundaries in the second dimension in which the business will operate. Are we to serve industrial customers only or industrial and consumers both? Are we going to operate locally, nationally, internationally, or intergalactically? The product decision and the market decision are tied together in some important ways. It is difficult to aspire to selling newspapers globally (a product). However, some have done very well distributing news globally (a service).

Positioning

This is the most crucial decision of all. In the broad sweep of business history, many organizations have successfully rethought their mission in terms of product and served market. It is more difficult, however, to reposition a company *within* their served market. It can be done. But it takes an enormous amount of energy to do it. A company may commit to differentiating itself as a high-quality producer. It had better think through carefully what is meant by "high quality." It may position itself as a service-oriented company. Its products are solid but not necessarily the latest and the greatest from a technological point of view. But its customer service is superb. So its customer service is the product even though it appears to sell a physical product. Another company may position itself as a low-cost producer of solid, but fundamental products. Another might be fashion oriented.

The mission statement should address all these dimensions in a concise and "pursuable" way. The temptation always exists to define mission very broadly so as not to foreclose on opportunities that may arise later. This is commendable and shows a proper appreciation for the future, but it may be so broad as to fail to provide a focus for the organization. If your mission is "to provide manufactured goods and services to industrial and individual consumers in the global marketplace," you probably haven't been very helpful to the long-range planners. On the other hand, if your mission is "to provide round-head, slot-drive, 6-32 brass machine screws in lengths between ¼ and ¾ inches to hardware stores in Keokuk, Iowa," you probably don't need long-range planners.

The Strategy

The strategy is a definition of how we intend to accomplish the mission. It is broad and directional in nature. It gives policy direction for the conduct of day-to-day affairs. It provides a litmus paper test for the short-range plans of the organization as well as the day-to-day ad hoc decisions that are made at every level of the organization, mostly out of sight of the corporate leaders. It may be incorporated within the mission statement, appended to it, or exist separately. This is mostly a matter of taste.

One company's mission statement is "to provide highly engineered polymer products to the worldwide automotive industry," a very concise and complete mission statement. Its strategy is "to differentiate [itself] by providing products and services which are judged superior [by the customer] in meeting the needs and the expectations of the customer and the ultimate consumer." Note that the strategy doesn't stand by itself. That strategy by itself could be that of McDonald's as well as of an automotive supplier. But in combination with the mission statement, it gives focus to the direction of the company.

Values

Every corporation should have a well-thought-out set of values. "Values" as used here refers to the ethical burden that the enterprise accepts as its legitimate cost of doing business. These should be carefully considered and entered into, more like a marriage than a business deal. If the list is an amalgam of law, union contracts, and public relations then it is not a list of values. Values are things you would subscribe to if there were no laws and union contracts and public relations wasn't important. Granted it may be the same list, but the organization can detect phoniness here at the level of one part per billion. Preach values that you don't really believe in and the world will discount everything else you have to say. Total silence is better than sanctimonious tripe. Should the leader choose to deal with values he should consider his *obligations* to employees, shareholders, communities, the environment, social issues, and the future. Speak when you feel strongly. Otherwise remain silent. The strongest value statement the author ever heard, and it was from a fine leader, was in response to a question from a subordinate about a risky course of action to be taken with a difficult customer. It was to the effect of "Why share this information? It can be used against us." The answer? "Because it's the right thing to do." It was, we did, and it set the whole organization on a new, higher ethical plane. Values often set the priorities for the operations when ugly decisions have to be made in day-to-day operations. Thus they extend the strategy and become tactical "rules of engagement."

Priorities

Values and priorities are easily confused. This is because one's values, to a great degree, set one's priorities. But "priorities" as used here has a somewhat different meaning. A good place for the executive to start is to ask himself, "What are the few things we must do superbly if we are to accomplish our mission?" In that sense they can also be considered strategic objectives. Not all the work of the organization is critical, thank goodness!

Most organizations have too many priorities. Often this is a tipoff that they haven't really thought through their strategy. There is also a tendency for managements to say that all these priorities are of equal weight, and none is subordinate to the other. Perhaps this thought gave rise to the classic obfuscation "first among equals," an idea that has never made sense in this application. How can anyone handle 19 different values simultaneously, especially when they are all of equal weight in the decision-making matrix? If the employees really believes this, then management has put them in a classic state of decision-making gridlock. To refuse to set out the priority order is a management copout. It says, "I can't tell you in advance which values are trump, but I will know it when I see it, and you in the organization are responsible to assure that I like it when I review it."

Priorities should be few. They should also be prioritized with respect to one another. Key terms should be defined carefully. When management uses words like quality, customer, employee involvement, and productivity, to name a few, they are obligated to define their terms. Organizations and the people in them have a way of using the same words to mean different things without even realizing it. Sometimes the same word takes on different meanings when used in different situations. This is another way of ducking the sometimes difficult circumstances caused by having values. It is also a source of confusion for the entire organization.

The mission statement and strategy quoted earlier is accompanied by the following set of priorities;

1. Employee safety
2. Product and service quality
3. Elimination of waste in all business and manufacturing processes

Clearly, the first is a value that has also been made a priority. The second and third are obviously priorities linked to the strategy and execution of the mission. Why make this one value a priority? In short, the executive believes that employee safety is an overriding value. Moreover it is one that can easily be compromised in the day-to-day pursuit of the others. He has made it clear that this is not to happen. Ask him and he will ask in return, "What priority would you suggest I put ahead of employee safety?" Beware of your answer!

Formulating the Vision

Vision is defined in part as "unusual competence in discernment or perception; intelligent foresight." This is the sense in which it is being used here. Vision in this sense would seem to be a faculty of individuals rather than of organizations or committees. Visioning and business planning are not the same thing. For an individual's vision to be effective, he must be at the top or at least in a position of great influence within the organization. He discerns things in the present that have profound implications for the future. Or perhaps he perceives things in the future that have profound implications for the present. It can work either way.

The visionary can charter committees to study particular aspects of the present or the probable future. The visionary will almost certainly need to pick the brains of many people. The visionary sucks up input like a whale sucks up krill. He takes in massive amounts of material, some valuable, much not. He filters and assimilates and the vision takes shape. He develops a unique insight into what must be done in the near term to be positioned for the future, or perhaps what must be done in the near term to create a different future.

Clearly, mission, values, strategy, and strategic objectives are not developed in series. This is not a tidy process of reaching a conclusion on one subject and then going on to the next. This is a convoluted subject that involves plenty of writhing and looping back. It is wise to reserve publication of the first element until the whole job is done. Otherwise the leader may find himself stuck with either a mission statement that doesn't quite fit or the need to republish a modified version before the ink is dry on the first one. The implications for organizational commitment are obvious.

Securing Commitment

The first and most important thing to be said about securing commitment to the vision is that it is a lot easier to talk about than it is to do, and it's not that easy to talk about. Securing commitment requires several things. The vision must be communicable. It must be credible. It must be relevant. It must be worthy.

Communication

For the sake of expression let's take this whole body of thought—the mission, strategy, the values, and the priorities—and call them collectively "the vision." No matter how elegant and fine-tuned this vision is in the mind of the leader it must be capable of being packaged and communicated in terms that can be related to by the maximum number of people within the organization. If the language and the terms of the statement can only be understood by the lawyers or accountants, if the statement requires a Ph.D. in English Literature to understand it, if its meaning hinges on the use of obscure words or presumes arcane knowledge, then it is not communicable and the leader cannot expect it to have much influence on the organization. But this is also a litmus paper test of sorts. If the leader cannot express his vision in straightforward terms to others, then it follows that they haven't thought it through sufficiently to bring the pieces into focus in their own minds. Focus is what these statements are all about.

This difficulty is often perceived by the authors of the vision statements. Unfortunately, it too often follows that this difficulty is sidestepped by resorting to slogans, such as the brief, punchy statements that one might expect to see broadcast between rounds on Friday night boxing. The three-word slogan seems to be the treatment of preference in this school of thought. This approach seriously underestimates both the intelligence and the sensitivities of the audience. Most people have a built-in, highly sensitive smoke detector. They can smell this kind of phoniness a mile away. It not only fails to turn them on, it turns them off.

The vision needs to be expressed in terms that bring it into focus for as many members of the organization as possible. The pieces should fit together such that the whole is somehow greater than the parts. It should be simple without being simplistic. Again, focus is what these statements are all about.

The best way to achieve this degree of clarity is also the most painful. The leader must write it himself. The best way to lose a potentially vivid and effective statement of the leader's vision is to turn it over to a "wordsmith." If he can express it clearly to a writer, then he can express it on paper himself, in words that those near him will recognize as his own. This is another remarkable faculty of the built-in smoke detectors. They sense with remarkable accuracy "that isn't him talking." Third-party editors may polish and shine. After all, that's what they are paid for. But they are always tempted to put a bit of themselves into the work. They're also paid for that to a degree. Finally, they are required to work through any fuzziness that they might have in their own minds as a result of the visionary's failure to communicate with *them* in clear and concise terms.

Credibility

The vision has to be believable. The customers for the vision, the members of the organization, must react with "aha," not "what?" The vision must fit the culture and the paradigms of the community, and not be foreign to it. This can be particularly difficult if the vision is expected to move the organization to a new way of behaving that is radically different from the former one. This may call for a great deal of mind preparation on the part of the executive before it is introduced if it is to be credible. Paradigms being what they are, this is apt to be a blind spot for the leader. The vision may well be crystal clear and totally credible to the leader from his point of view. But if the vision is, as it often is, the result of a paradigm shift on the leader's part, then it might not be so credible from the paradigm of the organization. That's why it is so necessary for the leader to be in touch with the community, if for no other reason than that he must give credibility to his vision in their minds.

The most difficult and important task for the executive in establishing credibility for the vision is to discover a latch point in the paradigms of the organization to which he can attach the new vision, so that the organization will simply believe in the truth of the vision. Then and only then can the executive lead them onto what may be a different path. In short, it is necessary to establish credibility first on the terms of the organization before they can be led into a paradigm shift on their own.

Relevance

If credibility has to do with believability, then relevance has to do with pertinence. The common reaction may be "Okay, I believe what you say, but what does that have to do with me?" If this is the case, the vision is stillborn. Chester Barnard observed that leadership has to do with creating faith: faith in the leader, faith in common purpose, and faith in the *relevance* of the task at hand. Relevance exists when the organization senses that the vision has something to do with them and where they must go.

The second element that must be present is the confidence on the part of the individuals that they, personally, can make a difference. The leader can't tell them this. The leader must show them or, better yet, help them to show themselves. This is one of the underappreciated aspects of a pilot project. It helps to convince people that they can make a difference in "this" way, whatever it might be. If they can make a difference individually, then it follows that they can make a bigger difference when they join together in small groups with common objectives. This is the basis for all successful small-group activity. But if they can make a big difference when working in small groups, think of the difference they can make if they work together as a whole team! As this realization evolves, the organization becomes an excited, productive, animated community. Jaworksy might say at this point that the leader has recontextualized the situation and empowered the people. And to that this author would say, "Right on!"

Worthiness

Let us accept that the vision is found credible and relevant. It must also be worthy of sacrifice. Put simply, something that is worth sticking one's neck out for. At the risk of redundancy, the objective must be worthy of the commitment that the leader is asking of the individuals. Organizations do not follow into areas of risk, and we *are* talking about risk. Change is always risky in some dimension. Only individuals follow. First one, then another, then several follow those who preceded them, then the mass follows the growing tide. No one wants to be left behind. The community paradigm is shifting and no one wants to become the cultural outcast. But make no mistake, this is not an organizational or community decision. That decision is made person by person. For that decision to be made the individual must see that the objective is worth the risk. One might take small risks and make small exertions for small objectives. Unfortunately, small risks and exertions make small differences. Causing people to make great exertions requires great and noble challenges. To reiterate, no soldier ever charged a machine gun emplacement because it was in his annual objectives. For his family, his comrades, his flag, and his own ultimate survival he might. Great heroes are made of great causes. Creating such causes is the burden of leadership.

CHAPTER 5

The Improvement-Driven Organization

JOURNAL: The kid was right. It seems like a long time ago that we were on that plane together. A lot of water has gone over the dam. I wasn't able to get his words out of my head. As I talked to people I found myself discussing and listening in terms of that vision that he was urging me on about. As I asked ''vision'' kinds of questions and somewhat tentatively put out visionary kinds of ideas I sensed a response in the people, my fellow-employees, as the kid would say. Sometimes I felt like I was standing on top of a volcano. The earth seemed to rumble a bit. I got the idea that there is a lot of power down there just waiting to be tapped. A bit frightening actually. It occurred to me that this is the kind of stuff that revolutions are made of. Some lunatic senses all of that energy and finds a way to tap into it and changes the course of history. Then it dawned on me that not all revolution-aries are lunatics. That's just the way we ''establishment types'' always seem to think of those who would change the shape of things. Guess that makes me a flaming-eyed revolutionary. Now that I understand that, I feel a little more comfortable—I think. Now I can focus on how to tap that energy.

I've pretty well shaped my vision for the company. I bet I have written and rewritten the mission statement a dozen times. Started to give it to our communications person. Awfully good with words, but I couldn't bring myself to do it. Too personal, I guess. Thank goodness, it's getting shorter as I work on it. I'm pretty close now. I think I've got a handle on the values too. It's simply a matter of thinking through what kind of behavior would make me proud of this organization. Kind of a cue question, but it worked for me.

The tough part is the implementing strategy. What is the route for getting there from here? My marketeer recommended a strategy of increased market share. My chief financial officer recommended increasing return on investment. I can just see the whole organization charging up the hill for increased market share or ROI! The quality guy recommended quality as a strategic theme, of course. Not too bad depending on how you define quality. But I'm looking for something else. Something that ties it all together in a bundle that I don't have to constantly explain. Something that we can all relate to and contribute to. That's the only way we'll get everyone into the act. Getting everyone in is absolutely essential. For one thing I have become convinced that some very outstanding thinking comes from some very unexpected places. If they're not all in, how can you expect to get it all out?

Quality is what it's all about, of course. But not just product quality. Somehow the idea of creating product quality in a vacuum isn't realistic. Then it occurred to me. Sort of a mini paradigm shift I guess. I'd been thinking about it a lot while going back and forth to the office and during idle moments. Then all of a sudden, there it was. The idea of a mediocre organization creating excellent products is silly. The idea of an excellent organization creating mediocre products is inconceivable. The central idea is organizational excellence. Excellence in everything we do. Now, how does organizational excellence express itself? What is the central, driving characteristic of the excellent organization? What is the appeal that the excellent organization will respond to? If only I can tap into that!

Defining the Strategic Objective

The fundamental question to be addressed here is this: How does the organization wish to differentiate itself from all its competitors? A mission without a strategy is a pipe dream. And a mission statement without a defined strategic objective is nothing more than a slogan.

In a quality-driven company the strategic objective may sound something like this: Our strategic objective is to differentiate ourselves by consistently providing products and services that are judged superior in satisfying the needs and expectations of our customers and the ultimate consumer.

The Tactical Plan

At this point we have defined the mission and the values we hold for the organization. The mission tells us what we want to be. The values define the price we are willing to pay to get there, the means that are justified by these ends, the constraints that we accept on our behavior, the organizational ethics. We have also defined our strategic objective: how we wish to be thought of in the marketplace, how we wish to differentiate ourselves from our competitors. The question that remains is how to achieve and maintain that differentiation. For this question there may be only one right answer!

One would have to be a Rip Van Winkle to fail to notice that the world is in a constant state of change. If the world is constantly changing and the individual is not, it follows that a sizable gap will develop between the individual and the world in which he lives. There is a word for that gap. The word is obsolescence!

Consider people who come to work each day and then go home to their television sets never exposing themselves to anything more stimulating than reruns of "Gilligan's Island." They never learn anything new. They never even think about anything new. How are such people described by their peers? "Twenty years behind the times," "out of touch with the world around them," "not in the real world," and so forth—perfect everyday metaphors for describing the gap that has opened up between them and the world in which they live, for defining the nature and magnitude of their obsolescence.

The same thing happens to companies and sometimes whole industries. New needs are constantly appearing in the marketplace. Marketplace needs are business opportunities. To the first to perceive these needs and satisfy them go the spoils of legitimate competition. These needs, formerly unperceived, but now addressed, move osmotically from the needs into the expectations of the marketplace. The company that first perceives and then satisfies unfulfilled needs is not *satisfying* the expectations of the marketplace, it is *creating* them. Now it rests with the competitors to satisfy, if they can, the newly created expectation.

New processes and technologies are constantly becoming available to the producer. New materials are being developed, new substitutable products are being made available by new competitors. People with new skills are being produced by our educational institutions. Companies that are too complacent, too lazy, or too blind to recognize the opportunities that these changes represent and keep pace with them, or rather to exploit them, quickly become obsolete.

Some would describe this process as constantly shooting at a moving target. A useful metaphor, but a bit too reactionary for the company that would differentiate itself. That company is not shooting at a moving target. It is moving the target that others have to shoot at. Shooting at a moving target suggests a company that is constantly reacting to changes in the world and therefore one that is constantly just about a step behind. Depending on the company's strategy, this approach may serve well. If by being just a bit slower the product or service is also a bit better or a bit less expensive, the company may be able to exploit the opportunities that others open up. But what about the company that is the established leader? What does it sense and respond to in order to maintain the leadership? It is obvious that the company that senses where the world is shifting to and responds with a corresponding adjustment will rarely be out in front, for the simple reason that the changes the company is responding to did not just happen, but were created by someone, probably a competitor. The world-class company is not driven by continuous adjustment, but by continuous improvement.

It is probably also true that internally the company is not so much product driven, as process driven. It is fully aware that it is products and services that it delivers into the marketplace, but equally aware that those products and services are the inevitable result of the processes used to create and deliver them. Therefore, the company is constantly working to improve the processes it uses to sense the needs and the expectations of its customers and the ultimate consumers. It is constantly striving to reduce the amount of time and money required to respond to those needs. It is constantly honing its processes to improve the capability to produce products with ever-diminishing variation from item to item. It is constantly working to reduce the cost and improve the responsiveness of its internal processes. It recognizes the long-term imperatives and doesn't wait to be forced to action by competitors or by legislation or by community pressure. It recognizes that responding under these conditions is almost always more expensive and less satisfactory than a more deliberate, longer-term response. The world-class organization is driven by continuous improvement in all it does—continuous improvement, pure and simple.

Continuous improvement requires the enlightened participation of the entire organization. Moreover, continuous improvement, which is a process in its own right, tends to be a rather messy, chaotic sort of activity that appears anything but orderly when viewed from the outside. Continuous improvement–driven organizations resemble not so much armies moving forward in efficient lockstep, but rather ant colonies. In observing ants at work one would be hard pressed to perceive any overall plan or guiding intelligence. However, one definitely gets the idea that each ant knows exactly what it is doing and is proceeding with confidence in the value of its contribution to the colony. That is the look and the feel of the organization driven by continuous improvement.

But there are some things the leader must understand before he can effectively implement and maintain an atmosphere of continuous improvement within the organization.

First, he must understand that continuous improvement is an unnatural state among communities of people. Newtonian principles apply here. Systems, including social systems, at rest tend to remain at rest. This is the antithesis of continuous improvement. Even individuals who pursue continuous personal improvement are viewed by the community to be a bit odd, and they are. Necessity, as reported, is the mother of invention. Societies that live in undemanding environments tend to remain in a quasi-primitive state.

Communities at rest tend to drift. Some of that drift may be in an upward or improving direction. But much, if not most, of that drift will be in indifferent directions or in negative, degenerative directions. If this is true, then the principal challenge of the leader is to maintain a state of continuous necessity, thus driving continuous improvement.

This idea in turn raises the question of the shape of the continuous improvement curve. Some would claim that it is linear and upward climbing, as shown in Figure 5.1. If one's scale of measurement is gross enough, the curve may appear this way. If we look at society at large over a long enough period of time, the curve may look like this. That's the effect of the central-tendency theorem. But organization is not society at large and the time scale on which it operates is not that long.

Others, paying homage to that cynical-sounding law of diminishing marginal return, might assert that the shape of the curve is asymptotic, forever approaching some upper limit but never quite reaching it, as in Figure 5.2. Collective world history would seem to deny this. No objective evidence supports the idea that mankind is approaching some limit in this way. Companies do, but these are the dying companies. We speak of mature industries, those in which the shape of the growth curve is asymp-

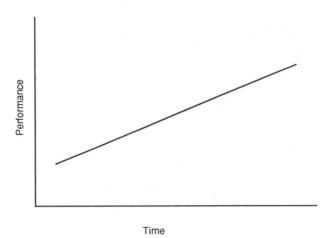

Figure 5.1   *Linear improvement curve.*

totic. The more optimistic assert that there are no mature industries, only mature companies. The truth may lie somewhere between, but the fact is that this is the growth curve of a dying company. The implication is that there is some natural performance limit beyond which the company cannot grow. But that is not the continuously growing company.

The shape of the improvement curve in the continually improving

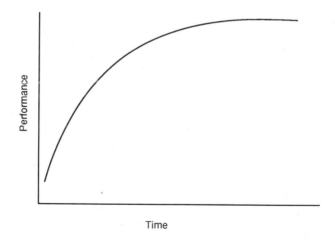

Figure 5.2   *Asymptotic improvement curve.*

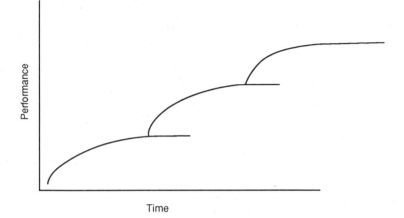

Figure 5.3   *Actual improvement curve.*

company is one of asymptotic curves stacked one on top of the other, as shown in Figure 5.3. It combines the elements of both the linear upward curve, since that becomes its long-term shape, and the asymptotic curve, which is its short-term shape, into a different form. Breakthrough occurs, improvement accumulates rapidly. As the juice is extracted from the breakthrough event, the curve begins to flatten out, until the next asymptotic improvement is launched from the newer, higher level of performance. The leader must recognize that these points where new curves are launched from old ones are discontinuities in the normal state of nature. They are breakthrough points, where the rules of the game have been changed. What brings them about? That is the work of the leader. It is the leader who is never satisfied with the status quo. The leader sees both threat and opportunity before the others in the organization. It is the leader who provides the vision that creates other leaders within the organization and empowers them to realize their own improvement objectives. It is the leader who causes these purposeful discontinuities in the system which we bundle together into the idea of continuous improvement.

CHAPTER 6

The Continuous Improvement Model

*JOURNAL: This is frustrating work. Every time you figure out how to remove one rock you find two or three rocks under it. But we **are** getting there, one rock at a time, I guess. It's clear to me now that the objective of our strategy must be continuous improvement. That also seems to be the common characteristic of the world-class companies I've been looking at. It's why they are leading companies year after year. They're restless in this way. If we continuously improve, **and** at a faster rate than our leading competitors, then we must eventually overtake them. This is a distance race, not a sprint. But continuous improvement is the objective of our strategy to achieve our mission. What are the details of that strategy? A strategic objective is not a strategy. Here I go turning over rocks again. What are the areas in which we must continually improve; everything? How does one go about continuously improving? Are particular skills involved?*

One thing seems sure. People are beginning to move around a little bit. Some are beginning to come forward with new ideas. A few are experimenting. This thing may be starting to catch on.

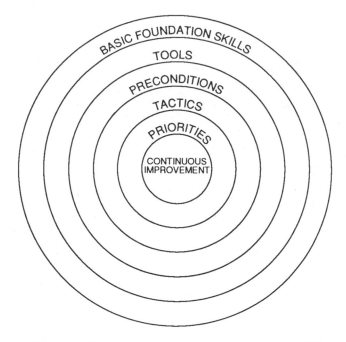

Figure 6.1   *Generic Continuous Improvement Model*

The Generic Continuous Improvement Model

The continuous improvement model can be thought of as a group of concentric circles, a target as it were. At its center is the ultimate objective, the bull's-eye, continuous improvement. The surrounding rings represent the prerequisites that must be dealt with before a state of continuous improvement can be established and maintained. These are arranged more or less in order, from the most elemental at the outside, gaining focus as they move to the center. Figure 6.1 depicts the generic continuous improvement model.

Priorities

The ultimate objective we have selected is continuous improvement. But continuous improvement in what? It would be easy to say that continuous improvement in everything that we do is the objective. That seems to have a bit of surface validity. But it wouldn't be too helpful. Without any more focus than this it is easy to fall into the trap of engaging in organizational origami, creating pretty little things that don't really contribute much to the central issues of the company. It is necessary to identify those few critical areas where outstanding performance is required if we are to carry out our strategy and accomplish our mission. These are the organizational priorities.

If we truly distinguish ourselves in these areas, everything else will take care of itself. It is also helpful if these priorities are such that all employees can relate to them and see opportunities to contribute in their individual spheres. This is what leads the members of the organization to conclude that "I *can* make a difference."

Identifying these few, critical priorities for the organization also avoids the difficulties that management inflicts on itself when they identify, over time, nineteen different priorities, all number one and which come into frequent conflict with one another. No one can manage this kind of priority set. In the final analysis, it is not necessary to encumber the organization in this manner.

The continuous improvement model must be tailored to the needs of each organization. But for purposes of illustration let's assume we have decided that there are two areas in which outstanding performance is required if we are to execute our strategy. These are quality and the elimination of waste. Most organizations can probably subscribe to these. In addition, most will have one or two others they will wish to add. In our case let's add the value "employee safety"—not because it contributes in a big way to satisfying our strategy, but because it's the right thing to do, a moral obligation that we feel strongly about. Substituting these priorities for the generic "priorities," we have the model shown in Figure 6.2.

Safety

Safety is freedom from danger, risk, or injury. The workplace often has unusual potential for injury. Providing a safe workplace is part of the set of social responsibilities that naturally accrue to management. One might have a difficult time resisting the idea that it is first in that set. It is also true, although strictly secondary, that safety goes hand in hand with the achievement of quality. Most people with experience in industrial management will have observed the links between housekeeping, safety, and quality. They invariably move together. Accidents are also wasteful. They are expensive in dollar terms. The cost of insurance and workmen's

Figure 6.2   *Priorities specified.*

compensation, the loss of valuable skills on the job, a whole host of reasons make safety a productivity issue. But the foremost reason for giving safety a top priority is because it's the right thing to do.

Quality

It is absolutely necessary to define key terms. Perhaps the most common source of organizational confusion is the failure to take the time and effort to define in concrete terms what we mean when we use certain words. Typically, we use these expressions with full confidence that we all understand them the same way, when in fact we are all probably attaching different meanings to them, fuzzy meanings at that. Quality is prominent on the list of such words. The word "quality" as used in this book means: The degree to which products and services satisfy the needs and the expectations of the customer and the ultimate consumer.

You've surely heard simpler definitions of quality. But while simplicity is desirable, it's not the ultimate objective.

"The degree . . ." Quality is a continuous variable. That is a statistician's way of saying that it can taken on any value within a range of values. That range can be from zero to 100 or zero to infinity or minus infinity to plus infinity. Using the popular scale for classifying pulchritude, it can even be from zero to ten. And it doesn't have to be a discrete number. It's okay to decide that the object of your affection is neither a nine nor a ten, but rather a 9.9774. (After all, no one is perfect.) In the case of quality, this expression, "degree," puts forth the rather controversial notion that quality is *not* a "you have it or you don't" sort of characteristic. A statistician would say that quality, in this sense, is *not* an attribute. An attribute is a binomially distributed characteristic. Bi-nomial—two names. The product is broken or not broken, one of the two. There is no third alternative. It works or it doesn't. It conforms to its requirements or it doesn't. It has quality or it doesn't. We're talking about a system that classifies things into one of two cells. Thanks to the need for specifications, we in industry are sometimes led to thinking of quality as an attribute. It isn't. It is a continuous variable. Ask any consumer.

". . . products and services . . ." The quality of the product itself cannot be separated from the quality of the services with which it is delivered and, if appropriate, maintained. The assertion was boldly put forth earlier that customers don't buy products, but rather they buy services. The physical product is merely the package in which the service is delivered. That thought applies here. But more than that, the potential customer has a total buying experience. The convenient availability of the product, the manner in which it is packaged to be taken home, the simplicity of the assembly instructions, and the need for only commonly available tools are all service elements. And yes, even whether or not batteries are included. Add to this the convenience, courtesy, and responsiveness with which postdelivery service is provided and you have a total buying experience which will all be included in the customer's judgment of your performance.

". . . needs and expectations . . ." Customers have both needs and expectations. Satisfying customer expectations is not the ultimate measure of quality. It is the minimum cost of participation in the marketplace. In poker parlance, it's the ante. This is because consumers are conscious of their expectations. True, these may be assumed and therefore rest in the back of their minds, but fail to satisfy them and they come forward instantly. The company that fails to consistently satisfy customer expectations will not last long.

Needs, on the other hand, are not in the front of the customer's minds.

Perhaps they are not aware of the need. This was the legitimate basis for the consumer product safety act. It is not reasonable to assume that all parents, for example, are aware of the hazards involved in the ingestion of lead-based paints that might be used on children's toys, let alone how to recognize them. Therefore, one can either educate all consumers as to the risks and how to recognize lead-based paints, or laws can be passed that prohibit their use on children's toys.

In other cases the customer may have needs that he or she is aware of but that haven't migrated into the area of expectations because either no one knows how to satisfy them or no one has chosen to. There being no alternative, the consumer has no expectation. This latter case is common and particularly frustrating. The industry, in its collective paradigm, has decided that the market is price-driven to the exclusion of quality and all the players are competing to put out the cheapest product possible. The consumer is trained to expect nothing better simply because he has no alternative to the products being offered. All are the same, all are poor. Then someone offers a desirable product at a premium price. The guffaws of the traditional producers quickly strangle into sputtering when consumers move huge amounts of volume to the new suppliers. Granted, there remains a market for the inexpensive commodity product, but there is also room for the premium product.

The company that is satisfying needs is *creating* expectations in the minds of the customer. For a time, at least, theirs is the only product that can satisfy this new expectation. Companies whose only goal is to satisfy expectations will always be in pursuit of those that are creating them.

". . . customer and the ultimate consumer . . ." The customer may or may not be the ultimate consumer. Both must be satisfied. Few products go directly from manufacturer to consumer. Most go to an intermediate customer which incorporates them into its product. Or perhaps they go to a merchant who offers them for sale to the ultimate consumer. Satisfying the customer is the price of admission. Satisfying the consumer is the price of success.

It is in this sense that we have chosen quality as one of the critical priorities of the company.

Eliminating Waste

Eliminating waste captures the productivity idea. We could even use the word "productivity" except for the fact that the term has been so terribly prostituted over the years. To many people the term conjures up the notion of minimizing cost per unit output without regard for the merchantability of the product, let alone for its quality as we have defined it.

We will define waste as: Cost or time needlessly consumed in activities undertaken to meet the goals of the organization. There are the obvious sources of waste: scrap, rework, returned goods, the costs of warranties and settling customer claims. These are primary "quality costs."

But eliminating waste also involves other activities of the business—the cost of making travel arrangements twice, the costs of superfluous meetings, meetings canceled at the last minute or started late because someone did not arrive on time. It also includes the extra time taken to follow procedures that are unnecessarily complicated, when simpler, more direct methods are available. It includes the cost of activities that don't really have to be conducted in the first place. Ultimately, work-in-process inventory is waste during those periods when value is not being added. In short, waste is all costs that really don't have to exist in order to accomplish the goals of the business.

Businesses that have been around for a long time are almost invariably disadvantaged when compared to recently established competitors. This is because the systems and procedures of the enterprise have existed for a long time also. They were established using the technology and methods of the past. They have been updated over the years, in a patchwork sort of way. Ballpoint pens have been substituted for turkey quills and perhaps computers have been substituted for file cabinets. But the basic architecture of the system probably hasn't changed much. We have been patching new cloth onto old garments.

Archaeologists excavating beneath ancient cities built on hills frequently find that the hills are the remains of other cities, one built on the ruins of the previous one since time immemorial. Today's inhabitants expend a lot of energy trudging up and down hills. This is the way with the business systems of long-established companies. Their systems have been assembled from the rubble of previous systems, patching the pieces together with new materials but never going down to bedrock to begin anew. The new company, not so encumbered, builds efficiently on bedrock.

An excellent test that can be applied in offices as well as factories is the value-added test. Simply ask the question "Is what I am doing at this moment adding value to my work product or not?" If the answer is no, someone should look closely at the activity to see if it can be somehow eliminated. If the answer is yes, can it be streamlined to reduce the cost or time involved? Notice that unnecessary consumption of time is considered a waste. Product sitting on a pallet awaiting processing is accumulating inventory carrying costs and occupying space that could be put to more productive use. But it is also interfering with our satisfaction of customer needs and expectations in other ways.

Often we find ourselves quoting thirty-six-week lead times to customers for products that may need only a few hours of processing time. The rest of the time is spent in stockrooms or queues awaiting processing. But customers can't always predict their needs thirty-six weeks in advance. They need faster response (they might not expect it, however). Knowing that they might need product thirty-six weeks from now, they are tempted to put it on order fully expecting to cancel the order at the last minute. This is another source of waste for the entire system.

Faced with lead time increases, customers are forced to enter "just in case" orders. As these orders flow in, the backlog increases and lead times increase further, stimulating more "just in case" orders. At some point suppliers may be moved to expand capacity to meet the demands of their booming market. Then, as the moment of delivery approaches, the just-in-case orders are canceled. This has a shrinking effect on lead time, which causes other just-in-case orders to be canceled, and now the collapse begins. As lead times shrink because of canceled orders, more orders are canceled, and by the time that reality arrives it is found that the increase in backlog was totally synthetic and the supplier is stuck with expensive incremental plant capacity—all because long lead times caused everyone to order just in case. Truly, time is money and it bites in many insidious ways.

Tactics

What's next? If continuous improvement in safety, quality, and the elimination of waste are to be the priorities of the organization, then how are these goals to be achieved? In short, these priorities make certain business processes critical. These critical business processes are problem solving, process improvement, and process control. (See Figure 6.3.)

Problem Solving

A problem, virtually by definition, is a situation that cannot be ignored. Weeding the backyard is something that can probably be put off until a more convenient time. It's not a problem. Fixing the hemorrhage in the plumbing system can't be postponed. *That's* a problem.

Similarly, if an influential customer calls and tells you he has a serious

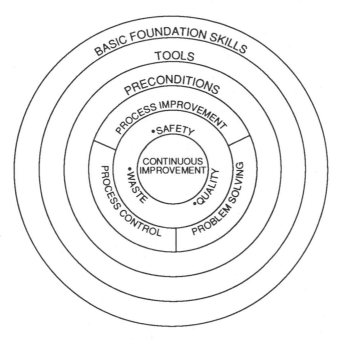

Figure 6.3 *Tactics specified.*

problem with your product, then you have a problem. It gets your full attention right now! The full resources of the organization are pressed into service to address the issue. It's an imperative. Corrective action is the name of the game. First, a word about corrective action. Corrective action is easily confused with remedial action to the detriment of the producer and customer alike. If, for example, the customer has complained about a shipment of product that contains a serious defect, that shipment of material must be replaced immediately. That replacement is not corrective action, it is remedial action. It gets the customer off of the hook and business starts to flow once more. This is properly the first priority. But corrective action is the action taken to identify and eliminate the root cause of the event for the future. How did this defect come to be created in the first place and how did it escape detection? We must identify the source of the defect so that it can be eliminated forever.

The corrective action process must be given serious attention by management. First, because it is so easy for the members of the organization to stop at remedial action coming up short of corrective action. This is the fire-fighting mode of operation. The complaint came in, the organization responded to the challenge, the offending material was replaced within twelve hours (a superb response), the organization heaves a collective sigh of relief, they congratulate each other on their "can do" spirit, and they return to work. Some organizations get so good at this form of "corrective action" that they solve the same problem over and over again year after year.

The second reason for management attention is that it is so easy to become confused in the area of root cause identification. It requires a high degree of perseverance to get to root causes. Unlike remedial action, which can create organizational legends and make heroes, corrective action is not romantic work. The individuals who practice it get to be a real nuisance to those who don't. The investigators must tenaciously question why? why? why? as they dig through layer after layer of superficial symptoms until they get to the bedrock of root cause. Then, having arrived, they must see the cause eliminated. These are the real heroes of the organization and they must be identified, encouraged, supported, and celebrated if continuous improvement is to become a reality.

Process Improvement

The processes we are speaking of are not just the technical, manufacturing processes. They include all the processes of the business, including managerial, administrative, technical, manufacturing, and supporting. In fact, in most cases, the manufacturing processes have been beaten to death over the years and the biggest rewards can often be achieved by improvement in the nonmanufacturing processes.

Process improvement is easily confused with problem solving, but they differ in some important ways. Problem solving is nonelective. The need is immediate. It probably requires a relatively narrow range of functional skills, but usually does require the participation of more than one person. Problem solving is difficult work and time is of the essence.

Process improvement, on the other hand, is elective. It usually requires that a broader set of functional skills be employed and time is usually available. Process improvement does not enjoy the benefit of urgency. While this would seem to be an advantage, as a practical matter there is nothing like a bit of urgency to create the climate for getting things done. Unfortunately, we become so busy doing the urgent that we fail to attend to the important. The improvement-driven organization doesn't make this mistake.

Being elective makes process improvement a spare-time job . . . and who has any spare time? Well, maybe it is more accurate to ask, "Who will admit to having any spare time?"

Process improvement also usually requires the participation of a number of people from a number of departments. All must be prepared to set aside time to work individually and collectively on a project that may not be terribly high on their individual priorities. And after all is said and done, much is said and little is done. After all, charity *does* begin at home! Process improvement is excrutiatingly difficult to maintain, yet absolutely essential to continuous improvement. If one of the three activities can be said to be more important than the others, process improvement would be it.

Maintaining a process improvement activity challenges the political dimensions of the organization at least as much as the technical or business dimensions. There are prerequisites to effective, sustainable process improvement.

The Sponsor

Each project must have a sponsor. To be an effective sponsor the individual must have the two P's: position and passion. Process improvement requires the application of resources—principally, people and their time. The sponsor must be in a position either to assign the resources and priorities directly or to influence those who can. Fortunately, because process improvement is elective, the project does not have to be a full-time job. It can be spread out a bit over time. But spread it out too far and interest will wane and the project will surely wither and die.

The sponsor must also have passion, passion to sustain the project during periods of competing interests in the sponsor's mind as well as others'. The project needs to be kept alive and moving at times when others, perhaps all others, simply wish it would go away. The sponsor doesn't have to lead the project personally. In fact, the principal job of the sponsor may be to manage the politics with the stakeholders.

The Stakeholders

The street definition of a stakeholder is someone who is in a position to mess up the project and might be so inclined. These projects require the participation of a number of departments and infringe on a lot of people's turf. Need more be said? Add to that the fact that most organizations optimize at the functional and departmental level, meaning that they suboptimize at the company level. Let's face it, most companies are run for the benefit of the functions, not the company. The purpose of process improvement is to change that and suboptimize at the functional level in order to optimize at the company level. This gives rise to all manner of political issues, which will have to be faced sooner or later. Sooner is better. Many a useful project has failed because the politics weren't addressed and were allowed to rear their heads at the eleventh hour and scuttle the implementation of a noble effort. The word "politics" is not used here in the pejorative sense. There are legitimate functional interests that have to be addressed. Let them be addressed up front. It is the job of the sponsor to identify the stakeholders, understand their needs, enlist their support, keep them informed, and generate a collective leadership for the project.

The Team Leader

A team leader must be selected. His will be the most time-consuming job. Therefore, it helps if the team leader is a member of the sponsor's organization. It is always tempting to put the company's foremost expert on the subject at hand in the position of team leader. This may be a mistake. The primary role of the team leader is that of facilitator and orchestrator. These are not universal skills and may not coexist with technical expertise in the mind and spirit of the resident expert.

Then there is the ever-present tendency of the resident expert to tell the team to just be quiet and listen to him. At this point his suggestion is either accepted because he is the expert or else the dogs of war are loosed. In either case the project loses. If the project truly needs his expertise, he may better serve as a member than as the leader. The leader must facilitate the meetings of the team and the efforts between the meetings. It is difficult, maybe impossible, to facilitate and participate at the same time. If the expert's participation is needed, perhaps he should not be asked also to facilitate.

The Team Members

Team members should be selected carefully. The team should be kept as small as possible consistent with the expertise needed and the appropriate representation of the stakeholders. The membership of the team need not remain static. It may evolve as the project progresses. But wholesale changes in membership should be avoided to prevent the "new team" from returning to zero and starting all over again, undoing the results of prior efforts.

Process Control

The objective of process control, be it statistical process control or otherwise, is to lock in the improvements achieved through problem-solving and process improvement efforts. People resist change. They will quickly migrate back to the old standards and the old ways of doing business if allowed. Improvement efforts lift the organization to new, higher levels of performance. It must not be allowed to slide back.

Statistical process control is often advertised as being an instrument of process improvement. Control is the prevention of change. Improvement is purposeful change. Control, then, is the antithesis of improvement. Statistical process control does, however, facilitate improvement. Processes must be put into a state of statistical control (more about that later) before they can be improved very much. Once operating in statistical control, the control charts begin generating intelligible signals about process performance. Responding to those signals intelligently leads to improvement, but that response is process improvement, not process control. Fundamentally, process controls operate to prevent change and therefore prevent improvement if taken to their logical conclusion.

Preconditions

If continuous improvement in safety, quality, and productivity is achieved through problem solving, process improvement, and process control, then what are the preconditions for these three critical activities? These preconditions have to do with the climate in which these activities are conducted. They relate to the organization, the reward system, and group skills. (See Figure 6.4.)

Organization

Continuous improvement, particularly in problem solving and process improvement, requires a great deal of group activity. One or more people will be asked to change their work patterns for the benefit of someone or something other than themselves. Seldom is a company organized in such a way as to encourage this kind of agreement. Our organizational structures are vertical, and usually functional.

The subcommunities of this organization are also functional units. The thing the members have most in common is their functional identity. They are in engineering or purchasing or manufacturing or sales. That's their

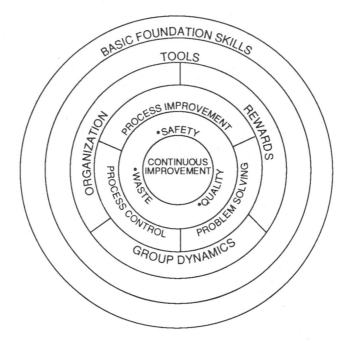

Figure 6.4   *Preconditions specified.*

focus. There's nothing wrong with this as far as it goes. But if our organizing efforts stop here, we will have institutionalized organizational isolation. Unfortunately, we usually stop here.

The products and services of the company do not flow vertically through the organization; they flow horizontally. But managers are hired to manage functions. The processes they optimize are functional processes. The economies they affect are functional economies. The processes to which they hold ultimate allegiance are functional processes. This invariably creates organizations with strong internal focus and little concern for the processes of other functions. "We do our job and they do theirs" becomes the order of the day.

It follows, then, that as the products of the collective organization flow horizontally through the system, they meet resistance, sometimes massive resistance, as they move across the interfaces from one function to the next. The structure of the organization itself gives rise to the "throw it over the transom" phenomenon in which each function does "its thing" to its standards and its timing and then places the resulting bundle, like an unwanted child, on the doorstep of the next function to do with it what they will. Needless to say, this creates frictions between functions which serve only to exacerbate the problem. "Engineering can be depended upon to give us products that can't be built." "Purchasing buys materials at the cheapest price; sometimes they're even usable." "Quality rejects everything they can and has no sensitivity to schedules." These allegations are often true because of functional fixation. The ultimate manifestation of this condition is found in consumer service situations in which the customer, usually unhappy, is informed that "it's company policy" or "I'm just following procedures." Both statements are perfectly true, both are totally irrelevant to the customer.

Matrix management was offered as a solution to this dilemma a few years ago. It was a terribly disappointing experiment in the experience of many companies. It should not have been. Matrix management is probably the oldest, most reliable form of organization known to man. Even the tribal chief was paralleled by the tribal shaman. Church and state have usually operated in parallel as two independent, but interrelated governments. Certainly every two-parent family unit (and they are conceded to be the best) is characterized by a matrix management structure. Although there are sometimes conflicts between the two dimensions of management, they are purposeful conflicts. In the mature organization, as in the mature family, they can be dealt with in the best interests of the family, not just the family member. Matrix management held the promise of providing the organization with the means to manage in both dimensions, horizontal as well as verticle. Perhaps part of the problem was that this style was touted by giving it a "high tech" name. Consultants and educators rushed to it as though it was something new and somehow magical. Like many subjects that are dealt with in that way, the truly difficult aspects were glossed over and the rewards presented with an almost spiritual aura. "Believe and it can be yours." Before the advent and after the demise of matrix management, we have had project management organizations within the framework of most companies. Ad hoc project teams that draw on the resources of the functions are still common. These are matrix management. So was the time-honored secretarial pool, still a pretty good idea in highly productive organizations. Accept the idea that if we cannot find a way to work in a matrix-like manner, we will never successfully implement a continuous improvement–driven organization, for the simple reason that optimizing the processes of the company requires the suboptimization of functional interests.

The failure of matrix management—and it didn't fail everywhere—was also due to the failure of management to address the reward system of the

organization. This remains the most important reason for near-universal
failure of managements that would change the culture of the organization.

Reward System

Management devises a reward system to encourage behavior that is
aligned with organizational goals and discourage behavior that is contrary
to those goals. This is very simple, straightforward, and effective. Why,
then, do managements that want to change organizational behavior fail to
recognize the need to modify the reward systems to align them with the
new behavior desired? Perhaps it is because the reward systems are so
deeply ground into the fiber of the organization that managements are
afraid to tamper with them. Perhaps it's because they feel that changing the
reward system within the company would somehow call into question the
way they are rewarded. Whatever the reason, there seems to be an almost
conspiracy-like agreement among senior managers to look over the top of
the reward issue all the time they are exhorting for cultural change.
Impossible! Rewards are so effective that they will always prevail. That's
why we have them.

Three kinds of rewards require scrutiny here: the monetary, centrality,
and recognition aspects. The question we must ask ourselves is "How do
we reward people for their participation in group improvement activities?"

Many companies have formal management-by-objectives programs.
These are purported to couple directly with the monetary rewards that the
manager and often key staff personnel will receive. The objectives are
almost always functional. They have to do with accomplishing specified
functional ends by specified points in time and at specified costs. These are
in addition to the perennial objectives of managing within a budget and
with often specified numbers of people.

The question is, Why would anyone in his right mind devote anything
other than token resources to the achievement of ends that may well
suboptimize the performance for which he will ultimately be held account-
able in order to execute a project that is near and dear to the heart of
another functional executive and for which the rewards, if any, will flow to
the other person?

Problem solving is urgent to the point of overriding these considerations. Annual objectives are often set aside in the interest of resolving truly
major problems. Those who decline to participate in these efforts or visibly
drag their feet are quickly written off by the organization. Besides problem
solving is a hero maker! Failure to perform to annual objectives is usually

forgivable in one who has conquered a major corporate problem. Problem solvers also enjoy increased centrality and prominent recognition. They have truly counted coup. They are the organization's Lone Rangers and Supermen.

But what about process improvement? One manager may have a project dear to his heart but how does he induce other managers to take an interest in it, particularly if it may compromise the internal efficiencies or structure of the others' empire? An informal, complex system of bartering for support of one another's projects is often the only alternative. It works, but only if the other manager has a project of his own that requires support. While this approach can work, it smacks too much of the congressional approach to making law. We know how well that works. It also makes for mischief. Knowing the difficulties and the perils of this method, most street-wise managers will shrink from accepting the responsibility for nonemergency process improvements. The senior executive should think very deliberately about how to change this dimension of the reward system to encourage not only the sponsorship of improvement projects, but participation in their success by stakeholders.

And what about team members? Why in the world would one agree, without substantial coercion, to participate in a process improvement team?

Given that one finds himself on such a team, he can perform in one of two ways. He can produce the kind of behavior that the powers that be desire to see exhibited or he can behave in a negative way. If he is smart enough to contribute to the work of the team, he is probably smart enough to disguise any noncontributory behavior sufficiently well to avoid organizational sanctions. This is often called going along for the ride without ever pulling the cart. Now there are both positive and negative consequences to each type of behavior. For example, it is common to find that when the model is worked out with the help of the people who have experience with this sort of team activity (not the bosses), the consequences for positive behavior are largely negative while the consequences for negative behavior are largely positive. Since feedback is either nonexistent or poor, there are no negative consequences to nonperformance. Anyone who really bends to the task at hand, which is desirable behavior, gets a lot of extra work. Seldom is anyone relieved of his regular duties. This translates into uncompensated overtime over a protracted period. If the project is successful, he can be sure he will be appointed to another team and will be able to do it all over again. The sponsor and maybe the team leader will get all the credit for success. If the project is less than successful, the team member will get his share of the credit for the failure and perhaps a bit more. If he just goes along for the ride, however, he is not going to expend an undue amount of uncompensated overtime and not likely to be asked to give above and beyond the call again soon. Management must think very deliberately about how to assure that the rewards for positive behavior are positive and the rewards for negative behavior are negative if these kinds of team efforts are to be successful.

Monetary rewards are fine and can often be worked into the recognition of team accomplishment, but one must be careful. Monetary rewards for ad hoc accomplishments can cause a great deal of dissension over who got how much, why, and how much was it worth? Ask almost anyone who has ever administered a monetary reward–based suggestion system.

Other types of nonmonetary rewards may be more appropriate. There is a great potential to reward the team and its members with centrality when executives meet with them, listen to them, and behave in a manner that recognizes the value of their expertise and their contribution. Centrality differs a bit from recognition in that it is behavior that brings the members of the team to the center of things, makes them members of a special, but informal club of sorts. They are not paid with money but with membership. And it is membership that can't be bought. Centrality sheds its light within the circle. Others know something important is going on in the huddle, but aren't sure what.

Recognition, on the other hand, is more a matter of shining a light on the team and its members in front of their peers. Recognition celebrates them and their achievement. This is a very powerful reward. Many a reluctant team member going along for the ride has been converted to a contributing member when the potential for centrality and recognition has been recognized. But negative rewards must also be present. Those who choose not to participate must be removed for the team and all must understand that their nonperformance was the reason. This serves as a positive reward for those who do contribute. It is a gesture of respect. It also serves as a negative consequence for those who don't.

Group Dynamics

Group activity involves the use of learned skills. There are facilitating skills be to learned by the leader. There are rules of participation to be learned by the team members. There is a base of analytical skills that all must share if all are to participate. Listening, communication, and presentation skills are important. These are skills that have been largely ignored by industry in their already too meager training activities. When quality circles were first introduced in North America, it was quickly discovered that supervisors needed to be backed up with facilitators if the circles were to be successful. This was a tacit confession that supervisors in our businesses and industries lacked facilitating skills. But what is the primary duty of a supervisor if not facilitation?

It has always been said (usually by us) that the United States is characterized by team sports. We are the global distributor of teamwork. The example breaks down upon even casual examination. Almost every team puts the player's name on the back of his jersey. This has been a selection criterion for some talented college and university athletes when choosing an institution of higher learning. No name, no play. Professional player contracts are negotiated individually. Players may find themselves on teams, but their interests are largely in "number one." Players on athletic teams change membership regularly and seem to have little lasting affinity for one team over the other. So do the employees of corporations. A baseball manager like Tommy Lasorda of the Los Angeles Dodgers, who, it is reported, bleeds Dodger blue when he cuts himself, is looked upon as a quaint anachronism. This is not the land of teamwork. We are seldom organized for it, rarely recognized for it, and more rarely trained for it.

Tools

If continuous improvement in some key areas is the objective, if problem solving, process control, and process improvement are the keys to this improvement, and if proper organization, focused reward systems, and training in group skills are the routes to achieving the necessary group involvement, then what are the specific tools involved? There are many. Decoding the necessary alphabet soup, statistical process control, design of experiments, quality function deployment, and continuous flow manufacturing are examples.

It is often assumed, without much thought, that these skills are somewhat universal. That is, all skills apply to all situations equally. Like many unchallenged assumptions, this one turns out to be somewhat unreliable when examined closely. Figure 6.5 catalogs some of the techniques that are currently being used or that should be considered.

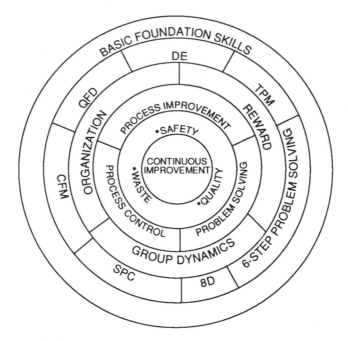

Figure 6.5   *Tools specified.*

Foundation Skills

This takes us to the final ring of the model. This entire structure rests on a commonly held, but often false assumption: that is, that the people in the organization at all levels bring with them certain skills, essential to playing out their respective roles on the team. These skills have presumably been imparted by the basic educational system and needn't be screened for. This is a false assumption.

People at all levels are presumed to be able to read and comprehend what they read. If people cannot read and comprehend, how can they understand our policies, implement our procedures, follow the job instructions, or behave safely in the workplace by following cautionary labels and instructions? How can they participate effectively in a team activity? Yet a large, growing percentage of the industrial population cannot read at a functional level. Possession of a high school diploma no longer assures possession of this basic literacy skill. Some companies are recognizing this problem and providing remedial opportunities to their employees who have this need. As commendable as this is, too many of those companies are undertaking this effort as a community service activity, not recognizing the essential, not optional, nature of reading comprehension to the needs of the company.

Similarly, basic math skills are becoming increasingly important to effective factory operations. We have spent enormous amounts of money to train factory operators and their supervisors in statistical process control and wondered why these efforts were without effect. One of the reasons is that the people being trained often lack the four basic math skills. These four math skills—adding, subtracting, multiplying, and dividing—are all that are required to set up and operate a statistical control chart. Only three are needed to operate a chart that someone else has set up. Yet these skills are lacking to a significant degree among the very people whom we expect to operate these processes in our factories. If these people cannot perform at this level of mathematical competence, how can they be expected to understand and accept the concepts that underlie these important production tools? Many managements have taken the position that understanding the underlying concepts is unimportant. "Just maintain the charts according to these rules, and take the prescribed actions when the signals printed on this card surface" (assuming they can read the card). This is a convenient copout, but a weak reed upon which to rest. Understanding the basic concepts is central to acceptance of and reliance on these tools. And acceptance and reliance are central to true implementation.

This educational deficit is not limited to the factory workforce. The reading comprehension skills of many technically trained college graduates are marginal once they get out of their own narrow fields. Writing skills among the technically trained are the subject of jokes—bad jokes. Composing a simple declarative sentence is an unwelcome challenge to many college graduates. Communicating complex ideas in writing is often beyond them.

The ability of many executives to communicate either in writing or orally is poor to nonexistent. Yet if managers and executives are to be leaders, they must be able to communicate effectively with those whom they would lead. As they move up in the ranks of the organization, the ability to affect that communication using the written word and public-speaking skills becomes essential. The great leaders of history can invariably be characterized by their ability to galvanize the minds of great numbers of followers using the written and the spoken word. And they invariably composed and delivered those words themselves.

These are not skills that we should assume. We should select for them, train for them, cultivate them, respect them, and reward them.

PART III

Economics of Quality

CHAPTER 7

Inspection Efficiency—An Oxymoron?

JOURNAL: I've decided that our quality manager isn't the intellectual cripple that some would have me believe he is. In fact, increasingly I find that he's a darned bright guy. I find myself watching him from a different point of view than I once did. First, he's a pretty adept survivor in an often foreign and hostile land. You've got to respect **that** *ability.*

In meetings I sometimes feel that I'm sitting on a different planet watching these people play out their roles in a game called "running the business." In a sense, that is just what I am doing. As I have become more and more preoccupied with changing the company I have become less involved in the day-to-day detail. I don't know whether that's good or bad. I admit that it scares me a bit, but one consequence is that I sit in these meetings observing what is going on rather than participating in what is going on. Later, after the meetings, I operate on what I have observed. A different style for me, but I think I'm accomplishing something by it.

When quality issues come up, I listen to the Quality Manager's comments and answers. As I look around the room, it's obvious that they don't always strike the other people as relevant to the issue at hand. But they make sense to me now. They didn't always. Maybe that's because most of us tend to look at issues from the inside out while he looks at them from the outside in. Both perspectives are needed, but as a group we don't seem to appreciate that fact. Something else for me to work on.

*When issues come up that aren't direct quality issues, it's as if he isn't in the room. But when he injects a thought, I find that it's usually a good one; something worth taking into consideration. The others tend to be impatiently polite, if he is quick about it, but they seem to feel that a quality guy can't do much but accept, reject, and talk in esoteric statistical code and quasi-religious-sounding philosophy. When there are "practical" decisions to be made, particularly if they're out of his field, it's assumed that he can't contribute much. He's treated a lot like a chaplain in a military outfit—out of the mainstream when it comes to the day-by-day military system, comes to the fore when bullets are flying and people are getting hurt, but somehow is thought of as having nothing to do with reality to the unhurt and unthreatened. I commented about this to him one day after a plant tour. I thought he might have been offended but instead he chuckled. He said he liked my analogy. I asked him why that didn't bother him. He said, "For the same reason that it doesn't bother the chaplain. I believe in what I'm doing. In my field also, we deal mostly in small victories. That doesn't make them unimportant. Then, it's not all that noble either. The occasional last laugh eases the frustration. But I **have** learned not to say, 'I told you so.' Instead I've tried to master the Mona Lisa smile; it's much more satisfying, and I don't get hit." That conversation has stuck with me. I'll admit that it bothers me.*

Later I followed up on a question that stuck in my mind during another of our endless meetings. I asked him why he kept talking about process capability when the issue before the group had to do with inspection escape. Sounded to me like he was trying to duck the issue. That isn't like him. He said, "Boss, you're confusing inspection efficiency with inspection effectiveness." I had to leave for another meeting but I've got to get back with him on that one. It went right over my head.

Objectives of Inspection

There are several possible objectives for conducting an inspection activity. Before instituting inspection in any process, it is necessary to examine management's motives. The nature of the inspection operation and even whether or not inspection is the answer hinges on the answer to these questions. The principle question is "What is management attempting to accomplish with this inspection?"

Remedial Inspection

Remedial inspection is inspection conducted to improve the quality level of the product exiting the process: in short, to weed out defects or defectives. Used in this sense a defect is an individual instance of unacceptability on a product. A single unit of product can possess more than one defect. A defective, used as a noun in this case, is a product exhibiting one or more defects. This distinction can be quite important, particularly in statistical applications.

If the objective is to weed out a certain defect, then the units of product will be inspected for that characteristic only, ignoring all others. If, on the other hand, the objective is to weed out defectives, the units of product will be inspected for all characteristics, rejecting any unit found to possess one or more defects (defective characteristics). This latter type of inspection is obviously more costly and usually of lesser reliability as a screening process.

If the stream of production is contaminated, then it contains an unacceptable level of defects or defectives. If the operators of the manufacturing process knew which units of products were defective or contained the defect, then they would have been set aside at the source. That not being the case, we know that every unit of production has an equal probability of being defective (an adjective here). Therefore, 100% inspection is in order. There is no alternative. Management will sometimes be troubled by the level of defects emanating from a process and institute inspection. But wanting to have their cake and eat it too, a normal management attitude, they will, in the interest of economy, insist on the use of a sampling plan. They will then form lots (meaning batches) of product and inspect them to this plan, accepting or rejecting each lot. Rejected lots are then screened.

Since the defects are more or less randomly distributed throughout the population, each of these lots can be expected to be more or less equally defective. Depending on the design of the sampling plan, and the luck of the draw of the sample, some of these lots will be accepted and some will be rejected. There is actually little difference, if any, between the lots accepted and those rejected. Other than removal of the defective pieces from the samples in the accepted lots (assuming that the number of defects allowed to be found in the sample of an acceptable lot is greater than zero), there is no rectifying effect on the accepted lots. They go into inventory just as defective as they were when they came in to inspection.

Rejected lots fare a little better (or worse, depending on your point of view.) They are 100% inspected, with all defectives found removed. Note in the previous sentence "all defectives *found*." Not all defects are found. One hundred percent inspection is not often 100% effective. Also, most sampling schemes are constructed on the convenient, if naïve, assumption that all defectives found in the lot are replaced with acceptable items. Although this makes the plan designer's arithmetic a lot simpler, it rarely happens in real life. The detected defects or defectives are removed and the remainder go forward.

Clearly this process has had *some* effect on the quality of the outgoing product. *On average*, the quality level has been improved. But considering that all the parts don't go into a big tumbler and get mixed together before going into inventory (regardless of what we think of the material handling function), this average quality level exists only in people's minds, not in reality. An interesting attribute of the average is that it is a statistic that doesn't necessarily exist in reality. Each lot of material now in inventory is either very good or unacceptably bad. What we have done is make the stream of production going to the next operation, or the customer, very inconsistent, lumpy, as it were. If management is really going to bite the bullet, then remedial inspection of this sort must be 100% inspection. The effectiveness of this procedure is dependent on several factors that will be considered later in this chapter.

Verifying Inspection

This is inspection for confirmation. It can be considered an audit type of inspection. We have substantial reason to believe that the process is capable and performing at an acceptable level of quality, meaning that any defects being created are well within the limit of acceptability. Being responsible managers, we find it prudent to verify this condition from time to time. However, it is at this point that many depart from the ranks of the responsible as well as the prudent. They will turn to the pages of MIL STD 105 to select a plan they feel they can afford to operate. After all, MIL STD 105 was invented by the government, so it must be efficient and effective (like the Postal Service, for instance). In most cases it is neither efficient nor effective. That's not the government's fault. They invented it for their purposes, not yours.

At this point the truly responsible and prudent manager will want several pieces of information. What is the true level of quality (percent defective) of the process being audited? What is the maximum level of defectiveness (percent defective) that can be considered acceptable? What risk of accepting a lot exceeding this limit are you willing to take? And what risk of rejecting a good lot are you willing to accept? Now you are ready to have your resident statistician design a plan to meet these requirements. It is really not that difficult. But this is where MIL STD 105 usually comes in. Management wants a sampling plan. They don't provide answers to the above questions. They usually get terribly impatient if pressed for answers. What they do know is how much they are willing to pay for the assurance—usually not much. The quality manager now usually turns to MIL STD 105 because he doesn't have the information with which to calculate a custom plan anyway. He selects the plan that requires a sample size he can cope with and that's what you get. What you usually get as a bonus is a plan that doesn't work very well. But in the meantime management is happy, thinking that they have this shield of protection and the quality person at least has a little temporary peace and quiet. When disaster strikes in the form of an ugly surprise to the effect that defective lots are escaping detection, management attacks the inspection function. Inspection attempts to defend itself, and manufacturing just keeps doing what it has been doing. Clearly, it is not their problem.

On the other hand, if management does answer the above questions, a plan responsive to the need can be designed. In this day and age it can almost be guaranteed that management won't like it. Attribute sampling plans like MIL STD 105 simply are not sensitive enough to satisfy today's needs in most cases. Custom plans can be designed but the sample sizes can be guaranteed to be unacceptably large for most operations. At this point some quality engineer may suggest a variables sampling plan. This means that instead of counting defects in the sample or simply classifying each item in the sample as defective or not defective, we must measure the characteristic, reduce the measurement to a number, a variable, and then process the resulting data to characterize the nature of the population from which it was drawn. Although statistically more efficient, it is also technically more difficult and a bit awkward administratively. In our unending quest for the easy solution, we press on.

The dilemma described above is what has led many of those with just enough statistical knowledge to be dangerous to proclaim loudly that the day of the sampling plan is past. The mere fact that you might be using sampling plans is prima facie evidence that you are statistically incompetent. In fact, with some customers, if you are using sampling plans, you will not be considered a source of supply. They are inherently evil. Statistical process control (SPC) is the only method to use! Nothing else is even remotely acceptable. But before we charge off in pursuit of this notion, let's examine the claims. There are many people with only a three-day course in SPC who think they invented it. They didn't.

An SPC scheme *is* a sampling plan! It's just that instead of forming lots, selecting a 75- or 100-piece random sample, for example, with which to make a decision, those who practice SPC form lots on a time basis. Selecting a small sample, rarely more than five consecutive pieces, they make a decision about the current performance of the process. SPC users may be deciding about a larger amount of production based on a very small nonrandom sample. So far lot-sampling plans would seem to be better.

There are differences that favor SPC schemes, however. First, the SPC schemes advocated are almost always variable sampling schemes while the lot-sampling plans they deplore, like MIL STD 105, are attribute schemes. As stated before, variables schemes make more statistically efficient use of the data. But there are variables-based lot acceptance schemes! Even the government has them. MIL STD 414 is such a scheme. People avoid it because of its apparent complexity. But it's not the plan that's complex, it's the use of variables data, which is inherent in both types of sampling, lot sampling and SPC. However, lot acceptance plans using variables will still require fairly formidable sample sizes compared to variables control charts. Is this where SPC has the advantage over lot acceptance plans? Now think! Is it reasonable to expect that five pieces taken from an hour's production, and that's a big sample for an SPC plan, can reveal more than thirty to fifty pieces taken from a lot of perhaps several pallets of production? Of course not! But SPC requires knowledge of certain aspects of the process performance that management tends to avoid creating in traditional lot-sampling situations. That's one of the reasons that they avoid variables sampling plans. They're "too much trouble." Statistical control charts won't operate unless you know the process average and have an estimate of process variability. Hopefully you also have some knowledge of the shape of the distribution of individual pieces. Given the same information, lot acceptance plans are very competitive with SPC. Now we are comparing apples to apples, not apples to Volkswagens. However, there are differences, some of which favor one form of control, some the other.

SPC charts can be maintained by the production worker. This gives the worker who must manage the process insight into its performance and the means to manage process centering and sometimes variability in near real time. Notice that I said, *near* real time. Some advocates point to SPC as giving real-time information while lot-sampling–derived information is derived after the fact. They observe that you cannot inspect the parts until after they have been made and that is too late. This is more than a bit of an exaggeration. SPC is also based on inspecting parts and that too is after the fact by whatever has been defined as the sampling period. Lot inspection does not have to be several days late and the results can be fed back to the operator. The real benefit of SPC is that, if the charts are maintained by the operator, he is involved in the quality effort in a real and tangible way. The information is coupled directly to the management of the process, and if it is process parameters that are being measured, lot-sampling techniques are inappropriate in any event. SPC has definite benefits and should be the first choice.

But lot sampling, particularly variables sampling, also has benefits. Using SPC, truly massive shifts in process centering or short-term variation may be detected on the first sample after they occur. But smaller shifts may take a number of sampling periods to be confirmed and may escape detection then, because the small sample sizes taken each time give rise to large sampling uncertainties. Overcoming these uncertainties requires the cumulative evidence of a number of samples to conclude with confidence that the process average has indeed shifted and requires adjustment. Assuming that the sampling interval was selected for a reason, seven or so intervals is a considerable delay in being able to make this decision. Four-hour checks are not uncommon, which means that it may be three and a half or more shifts of production before the process change can be acted upon. Some production runs do not last long enough for the shift to be detected and acted upon. A lot-sampling plan will characterize the lot of production so that an acceptance decision can be made with confidence. Because of the larger, randomly selected sample size, the resulting information can be furnished to manufacturing, sometimes sooner than SPC would provide it.

What about product audit activities? Ideally manufacturing is managing the process using SPC information, but management may wish to audit the process. At this point they can attempt to audit the SPC procedure to assure that samples are being taken as they should, properly evaluated, and acted upon. But this is a difficult and uncertain activity. The best way to audit the implementation of SPC may be to maintain a random lot-sampling audit scheme behind the process.

Also because of the small sample sizes, an SPC chart may tell the operator that the process has shifted either in average or variability and will indicate in what direction, but the chart will give only the broadest suggestion about how much. If the operator's job is to adjust the process back and the chart does not indicate how much, then he must approach the adjustment as a blind man approaches a curbstone, with small, incremental steps. This too takes time. A variable lot sampling approach, owing to its larger sample size, can provide a more useful estimate of the magnitude of the shift and hence the magnitude of the needed adjustment.

This is not an attempt to ridicule the use of SPC. SPC techniques are tremendously useful and accomplish some things that can be accomplished no other way. But that does not mean that other forms of sampling are obsolete and without value. One should compare the benefits and shortcomings of attribute sampling to those of attribute control charts, which are also scorned in some circles. Variables control charts and their benefits should be compared to variables sampling plans. The application should determine which is used. There is no universal solution.

In this chapter we have divided inspection into two types, according to purpose. One is rectifying, or sorting, inspection, intended to improve the quality of the stream of production. One hundred percent inspection is the only method to use for rectifying inspection. The other is verifying inspection, which could be considered audit inspection because it is based on the premise that the process has been producing acceptable product and we simply wish to objectively verify that it is continuing to do so. For audit inspection lot sampling is the tool of choice. This can be accomplished using either attribute or variables plans, but variables plans are rarely used, and this is unfortunate. They have tremendous power and need not be overwhelmingly difficult to use. This leaves SPC as the tool of choice in controlling processes statistically, an intriguing idea. That would make it principally a tool of the process managers, and that is manufacturing. This is SPC's best use. Confusion arises when the advocates of SPC recommend its use to the exclusion of other statistical methods and it is implemented and operated by the quality department as a quality tool, not a process management tool.

Inspection Efficiency

The notion of inspection efficiency is particularly relevant to rectifying inspection but is a factor any time units of product, be they from a sample or a population, are being inspected. "Inspection efficiency" as used here refers to the process average of the inspection activity, that is, the percentage of inspection decisions that are correctly made. This refers not to decisions about lots of material or processes, but to individual units inspected and about which a decision is made.

Consider that an inspector has just finished sorting a 1000-piece lot. There are fifty defective pieces in the lot, so that lot actually is 5% defective. The inspector correctly identifies forty of the defective pieces and removes them from the lot. But in addition, he misidentifies five acceptable pieces as being defective and removes them also. The inspector has made fifteen incorrect decisions out of the 1000 decisions made: the ten defective products he missed plus the five that were incorrectly identified as being defective. This means that the inspector made 985 correct decisions out of the 1000 decisions made, or 98.5% correct decisions. The inspector's efficiency is 98.5%. Some would say that this is a very good inspection performance. But when those ten defective parts are discovered, someone important is not apt to think so. After all, we are paying for 100% inspection and we expect as a result that there will be no defective products delivered out of that inspection. Right?

Before we have the inspector either promoted or fired, let's look at some of the alternative outcomes that were possible. Theoretically the inspector could have thrown away all the good product and accepted all the bad, with a resulting efficiency of zero. Most would agree that this is absurd and would not occur or, if it did, would be a failure in the system like switching the boxes into which the product was being sorted. Now let's get a little more realistic. Suppose the inspector was sound asleep while those parts were passing him on the line. When he awoke those 1000 parts had passed his station and none had been rejected. He would have missed all fifty of the defective parts but he would not have rejected the five good ones. His inspection efficiency under these circumstances would be 95%. By now that 98.5% doesn't look quite so formidable.

But if this is rectifying inspection, what we are really interested in is outgoing quality. The lot presented to the inspector contained fifty defective pieces of the 1000. It was 5% defective. After inspection it contains ten defective pieces. It is 1% defective. This difference represents the rectifying effect of inspection.

Inspection Effectiveness

Inspection effectiveness is the percentage of the defective product identified and removed by the inspection function. In our example, 80% of the defective items were identified and removed. Note that effectiveness takes no note of the good pieces that were misidentified. In the final analysis, the purpose of the inspection is identification and removal of defective product. While this thought might be bothersome, upon reflection it is of the real world. Inspectors get blamed for letting defective product escape. Seldom do they get in trouble for rejecting good product. This, for the simply reason that what they accept is passed on and any negative feedback occurs almost automatically. Rarely does anyone systematically look at the product rejected to determine if it is truly defective. It is obvious what happens as the world becomes more and more critical of defective products that are allowed to escape. The inspectors tighten their standards to avoid escaped defective products. But they cannot affect their inspection efficiency much, all other things being equal. In our case, the inspector is still going to make about 98.5% good decisions. In effect, the inspector is simply moving the window of error to the point of accepting fewer defectives at the expense of a greater number of false rejections.

Factors Affecting Inspection Efficiency/Effectiveness

Inspection efficiency and effectiveness are separate measures of inspection performance and both are important. They are also affected by factors outside of the personal performance of the inspector. These external factors must be recognized and dealt with.

Submitted Quality Level

The quality level of the product submitted to inspection has a profound affect on inspection effectiveness and usually inspection efficiency. Assume for a moment that acceptable products are not being falsely rejected. If there are no defective units in the production being submitted to inspection, then inspection efficiency will be 100% and so will inspection effectiveness (if you accept that zero divided by zero is 100%). But consider that there is one defective unit among the 1000. Depending on the nature of the defect, it could be extremely difficult to detect. Just staying awake until it passes by could be a challenge. If that defective unit passes by without being spotted, which is extremely likely, then the inspection efficiency is 99.9% and the effectiveness is zero. We have been highly efficient and completely ineffective.

If the production is, say, 90% defective, and inspection efficiency is also 90%, then efficiency will be judged to be pretty good by most standards. But remember, good product is being rejected also, so let's expand our example a bit. If inspection was 90% efficient, then the inspector made 900 good decisions out of the 1000. The 100 bad decisions were of two types:

bad product accepted and good product rejected. Let's say further that of the 100 bad decisions, eighty were to accept bad units and twenty were to reject good units. If 820 of the 900 bad units were detected, then the escape rate was 80/900 = 8.9%. Effectiveness was 91.1%; not bad under the circumstances. But the rectifying effect was not so good. We rejected 840 units of the 1000, having accepted eighty defective units and eighty of the 100 good ones. Of the 160 units accepted, eighty are defective. After all of this effort, we are shipping 50% defective product! (See Figure 7.1.)

Inspection Conditions

Rectifying inspection, regardless of who does it, is rarely designed into the job originally. It is usually hastily added when the need is apparent. When added, it is usually thought of as being a temporary expedient and therefore not deserving of much industrial engineering attention. It is not unusual for these temporary expedients to survive for years. Neither is it unusual to see people sitting on packing boxes in corners of warehouses under twenty-watt light bulbs sorting products into bushel baskets, conditions not entirely conducive to good decision making.

The factors affecting inspection efficiency, external to the job itself and the inherent skill of the inspector, are many. Light intensity and location is an obvious one. The color of the light when color is the characteristic being inspected is another. Inspecting on a moving conveyor line is another, particularly when defective products are being picked off and anything not picked off is treated as good. In these cases the inspector(s) must operate at the pace of the line and the density of the material flow down it. Extra time taken with one piece means another is not looked at. Clearly, tooling, jigs, and fixtures are another. Rarely are these inspection functions, particularly when considered temporary, provided with tools, templates, comparison samples, and so forth. Instructions and training may be meager. And finally, the characteristic being inspected for may lend itself to migrating mental pictures. "Too much," "too little," color shading, and "shades of gray" decisions, such as scratches and blemishes, are particularly difficult to referee and lend themselves to lowered levels of inspection efficiency.

*JOURNAL: Finally had that talk with the quality guy. He **did** know what he was talking about. He pointed out that his inspectors were rejecting about 25% of the products presented to them. The warehouse audit found about 5% defective in the warehouse after sorting. This is what everyone was excited about; the inspec-*

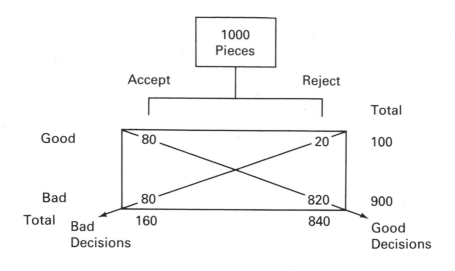

$$\text{Efficiency} = \frac{\text{Good decisions}}{\text{Total decisions}} = \frac{900}{1000} = 90\%$$

$$\text{Effectiveness} = \frac{\text{Bad rejected}}{\text{Bad in lot}} = \frac{820}{900} = 91\%$$

$$\text{Escape} = \frac{\text{Bad parts accepted}}{\text{Bad parts in lot}} = \frac{80}{900} = 8.9\%$$

$$\text{Percent defective shipped} = \frac{\text{Bad pieces accepted}}{\text{Total pieces accepted}} = \frac{80}{160} = 50.0\%$$

Figure 7.1 *Inspection effectiveness*

tors weren't doing their job. He had also audited the product being rejected by the inspectors. Found about 2% of the rejected product was good. This suggested to him that the most probable level of defectiveness being submitted to inspection is 28% defective. He had some interesting calculations in his shirt pocket. It boiled down to, out of every 1000 parts inspected, forty-three bad decisions are made. That makes the inspectors over 95% efficient. He sort of silently challenged me to complain about that. He added that they were catching 84% of the defective products, which he seemed to feel was pretty good under the circumstances. Then he described the circumstances. Another length of conveyor hastily added to the end of the manufacturing line, lousy lighting, inspectors working stooped over the conveyor, no tooling to help them in the sorting task, a rate of production that allowed 2.4 seconds on inspection time per piece (he had calculated it). Said he wasn't about to apologize for the inspectors. He was proud of them. I asked him why he hadn't presented this data at the meeting. Said that he had once before and had been accused of not being a team player. He was trying to make his point differently. The real problem is that you must improve process capability. I said, "You mean you **really can't** inspect quality into the product." He smiled and said, "Remember, boss, you said that. The last time I said that I was accused of dealing in clichés." But sometimes clichés get to be clichés because they are true.

CHAPTER 8

Structuring Quality Costs

*JOURNAL: Quality costs. A curious expression. My quality manager gets terribly excited about the subject. Curiously no one else seems to. I must confess that that includes me. I wonder why. We certainly are able to get excited about any other kind of cost. I called my quality guy the other day and asked him to come up and talk about it for a while. He showed up with an armload of tab runs and charts prepared to discuss the entire history of our quality cost experience. I'm afraid I disappointed him. Or at least I confused him. I just wanted to talk about quality costs: what they are, how they work, why I never seem to be able to get a straight answer to what I think are reasonable questions. I frankly don't know quite what I'm supposed to do with this information. I was still confused at the end of our conversation. I think he was too. He seemed to think that quality costs were important because somehow they translated his language, the language of quality, into my language, the language of dollars. I didn't think I was that narrow, but then maybe he's right to a degree. For sure, dollars didn't seem to be **his** language, not his native language anyway. We couldn't communicate in any way that I found meaningful. It's too bad. He's a good man and I know he is*

99

contributing a lot. At the end of the meeting I asked him for some reading material on the subject. I thought maybe I could figure it out for myself. He tells me that quality costs represent a large percentage of sales. He's right. But then, so do a lot of other things. Is he trying to tell me that I shouldn't have any quality costs? What's a good number? How do we compare to our competitors? What should the number be? But then maybe the important relationship isn't to sales at all.

The Nature of Quality Costs

It is important to understand what one is looking at when examining quality cost data. What a quality cost report is *not* is as important as what it is. A quality cost report is not a cost accounting report. To be sure, the report uses accounting data and is usually prepared by, or at least with the cooperation of, the accounting department. But for the data to be considered a cost accounting report it would have to be exhaustive, and no quality cost report deals with all the costs that could reasonably be called quality costs. And, all the information already appears in other reports. Also, no two people would agree completely on what data should be included or how it should be categorized.

A quality cost system is an economic model. Now, some of us are persuaded that there are as many economic models as there are economists. Someone once said, "If all the economists were laid end to end, they couldn't reach a conclusion." Well, not the same conclusion anyway. Each model seems to work well sometimes, none work well always. This is because the models do not include all the relevant factors and are not sensitive to all the possible interactions that can occur. There is an important defference between most economic models and quality cost models, however. The principal purpose of the models that economists design and run through their high-powered computers is to predict the future. Quality cost models are descriptive. They illuminate past and present cost relationships so that the manager can gain insight into the results of their strategies on quality and the cost of its achievement. They are trailing indicators.

That's why so many of the questions that the quality custodians get asked are impossible to answer or at least are answered with "It depends." It's also why so many people look the other way after a brief encounter with their first quality cost model. The executives first question is apt to be "How do our quality costs compare with those of our competitors?" The honest answer is "We don't know." For even if we had been able to extract the information from the competitor, we could be reasonably sure that his model would be different from ours. The next question would probably be "Well, what should the level of our quality costs be?" At this point the quality manager might revert to what could be thought of as a Sam Gompers answer. Sam Gompers was an early labor leader. During a particularly rancorous negotiation he was asked by a newspaper reporter what it was that labor wanted. Mr. Gompers replied, "More!" In a parallel vein, when asked what the level of our quality costs should be, an appropriate reply might be "Less!"—a legitimate, but admittedly unsatisfying response.

It is also not unusual for a manufacturing manager to look inside the system and see elements of cost that he has a difficult time associating with quality performance. At this point both the executive and the line manager are apt to turn their backs on the system and take up "more practical" activities.

As was said earlier, no one has a quality cost model that reflects the full scope of the impact of quality on the organization's performance. Also, being trailing indicators such models reflect in a muted way the impact of prior actions. How much they trail and under what circumstances is an important issue.

Organizing Quality Costs

Quality costs fall into two fundamental classes. First, there are discretionary costs, the expenditures that one can arbitrarily control by management decision. These are the independent variables in the management equation. The chief characteristic of expenditures in this class is that, being discretionary, they can be changed absolutely and in near real time. If one wishes to reduce these costs, people can be eliminated from the payroll and these costs go down immediately. One can add to these costs just as quickly. In this sense, discretionary costs are very responsive. This can be their greatest advantage (and their greatest curse). Attempts to respond directly to quality cost signals are apt to lead to disaster. Tell a line manager that the level of quality costs must come down, and quickly, and he will reduce those elements that are within his direct power to control. Those are apt to be exactly the wrong elements to be reduced for positive long-term effect.

The second class of quality costs are the consequential costs. That is, those costs incurred as a consequence of the strategy used in applying discretionary efforts. These are the dependent variables in the management equation. They are usually called failure costs, meaning they reflect the negative consequences of our inability to avoid the creation of scrap, rework, customer claims, and other sorts of effects.

Each of these classes of quality costs can be subdivided into cost categories. The traditional model would have one assign quality cost elements to one of four categories. There are reasons to consider expanding the traditional cost structure at the category level. An expanded quality cost model is offered in the following discussion. Structurally it looks like this:

Discretionary
    Prevention
    Appraisal
    Administrative
Consequential
    Internal failure
    External failure
    Process scrap

Categories Defined

Discretionary

Prevention: The cost of those activities carried out with the objective of preventing the creation of nonconforming products or events.

Appraisal: The cost of those activities carried out to assess the quality of the products of the organization.

Administration: The cost of those activities essential to the effective administration and management of the quality program which cannot realistically be categorized as either preventive or appraisal.

Consequential

Internal failure: The cost of correcting or replacing products that have been improperly rendered and have been detected prior to delivery into the distribution system.

External failure: The cost of correcting or replacing products improperly rendered and which were detected after delivery into the distribution system.

Process scrap: Scrap or waste inherent in the operation of processes using the current methods, materials, and equipment. Process scrap is usually considered unavoidable and is often referred to as "normal production scrap."

Note: The word "products" as used in the preceding definitions includes services. If services are the products of the organization or if they accompany physical products, they are no less products in their own right.

Collecting Quality Costs

Prevention Costs

Prevention costs are the costs of those activities conducted with the objective of preventing the creation of defects in the first place. Defining the elements to be included in this category can be controversial indeed. Because we are dealing with modeling, and not accounting, it is permissible for us to modify the model to suit our purposes. Rumors to the contrary notwithstanding, the traditional model did not come down from Mount Sinai graven in stone.

Prevention costs will almost without exception be the smallest of the categories. Perhaps for this reason it is often tempting to reach into all areas of the organization to identify costs to be collected into this category. The effect of overzealousness in this regard can be to make the preventive efforts appear to be greater and of more impact than they really are.

Also, most of the time spent on quality by functions other than the quality department have to be estimated or collected using secondary time-keeping systems. The fragments of time involved are usually small and involve a large number of people. The accuracy of these estimates is highly questionable and virtually unauditable. Also, accumulating this information is usually expensive if any attempt at reasonable accuracy is made.

Appraisal Costs

As previously defined, appraisal costs are the costs of those activities that we elect to perform to assess the quality of our products. This sometimes includes the performance of the processes that produce them. The principal appraisal cost is usually the cost of inspection. Inspection of purchased parts and materials upon receipt, in-process inspection performed by the quality control department, and final inspection and test are clearly appraisal costs. But deciding what to include in this category gets complicated as the inspection function becomes distributed into self-control activities and substitutes for 100% and lot acceptance inspection become more widespread.

It is recommended here that inspection that has been internalized into the manufacturing or service-producing activities be considered a part of the production operation and not treated as inspection for quality cost purposes. Admittedly this is controversial. The objective to some has been to make quality costs appear as large as possible in order to get management's attention. While this was an understandable strategy at one time, hopefully we are beyond that point in our respect for the quality issue and can take a more strategic view.

If manufacturing operators are responsible for inspecting their own work, then that effort has been included in the standards for their jobs. It has been made a part of the manufacturing task. Granted, it is still inspection in nature. But how can one consider that the part has been manufactured if the maker hasn't assured to his own satisfaction that the piece has been manufactured properly? Then what about the maintenance of process or product control charts? These clearly involve inspection work and often substitute for inspection. If the inspection and charting are performed by members of the inspection department, then they should be considered appraisal costs. If they are maintained by the production operators as a part of their responsibilities for product and process management, then they should not.

If collecting quality costs in this manner motivates management to internalize more and more of the quality management effort into the primary production processes, then it has achieved a desirable result. One should always consider the motivational effect of any performance report. There is no such thing as a neutral, motivationally unloaded management report. They are intended to stimulate response, and they do. In this case, the quality cost report can have the effect of motivating the organization to substitute in-process controls for traditional, after-the-fact, product inspection.

Administrative Costs

Many authorities recommend that certain administrative costs of the quality organization be categorized as preventive. For example, the argument is made that the function of the quality director's office and staff is really a preventive activity. It is certainly a legitimate quality cost, and clearly not an appraisal activity, so it becomes preventive almost by default. This kind of reasoning often leads to misleading quality cost reporting. One might question the effectiveness of the preventive program if the majority of the costs included in this category were those of the senior quality officer, his secretary, and other administrative activities. These activities, clearly marginal in their impact on preventing the creation of unacceptable products, can easily swamp what is already the smallest category of quality costs and create the illusion that there is more preventive activity than there really is. Hence the recommendation for inclusion of a separate quality cost category to contain these essential administrative costs.

Internal Failure Costs

Internal failure costs are the costs to remedy or replace products improperly rendered and detected prior to delivery into the distribution system. In addition to the obvious costs of manufacturing spoilage, scrap, and rework, we should consider including the premium portion of freight costs when the necessity for expedited shipping had its origins in a quality failure. The incremental costs of overtime and setup necessitated by quality problems and the need to break into production runs for makeup production are legitimate internal failure costs.

The impracticality of segregating these costs is presumed in most operations, but we should not be too quick to dismiss them. Often these costs are both substantial and, with a little effort, collectable. If the organization is currently going to extremes to collect discretionary costs of questionable strategic value, it might consider diverting that effort to the collection of these secondary internal failure costs, which usually have extreme strategic value.

Whether we choose to include such costs in the system or not is one matter. But we should still carefully think our way through the production process and at least identify the existence of these costs and be aware that they are there even if not included. Too frequently the architects of a quality cost model go to heroic lengths to collect discretionary costs while overlooking legitimate and collectable consequential costs.

In the area of manufacturing costs it is not uncommon to provide for an expected amount of scrap and rework in the manufacturing standards. In these instances many will insist that the quality costs include only variances from these standards. This is wrong-headed thinking. It's like playing golf using handicaps—okay for amateurs, but the pros play scratch.

Internal failure costs should include all scrap and rework whether or not it is provided for in production standards.

External Failure Costs

Obvious external failure costs are the expenses of warranty, claims and the other costs of dealing with premature product failure and customer dissatisfaction. In addition to these obvious costs, we should consider some not so obvious ones.

One example is the premiums paid on product liability insurance. The insurance coverages are maintained in anticipation of external failures. If the executive was confident that there would be no incidents, there would be no need for the insurance. Also, the premium on the coverage is usually based on claim experience and therefore is well indexed to past history.

The cost of travel and subsistence for employees who are visiting customers in the process of understanding and settling problems should also be included in this category. These costs are frequently picked up but misclassified. This work is often performed by quality engineers. Quality engineers are by default classified as a prevention cost in many systems even though the activities they are engaged in may be, for the most part, fire fighting. This inattention to detail leads to serious distortions in the quality cost model.

Round-trip freight costs for returned goods should be included. This is a big item for many operations that is easily overlooked.

Process Scrap Costs

Most people cry "foul" at the thought of including process scrap in a quality cost report. After all, these costs are unavoidable! We are talking about borings and shavings in a machine shop, edge trim in soft goods, the skeleton that is inevitable when circular shapes are cut out of rectangular sheets of material, flash pads from molded products, and startup and shutdown scrap from otherwise continuous processes.

Perhaps this reluctance is born of our collective impulse to look at a quality cost report as a performance report upon which we will be judged. This is unfortunate. A quality cost report should be thought of as a source of strategically useful information with which we can identify opportunities for process improvement. Therefore, we should adopt a "zero based" mentality in which all nonproductive costs are considered avoidable in the long run. It is clear, however, that these costs should be collected apart from the internal failure category. They are not spoilage costs. They are institutionalized waste and therefore opportunity costs.

Summary

Many of the ideas presented here are controversial. There is also plenty of room for argument over how particular activities should be classified. Some would argue that manufacturing's use of control charts is a preventive activity. Others might assert that product control charts are appraisal costs but process control charts are preventive. Material review is a formal process for dispositioning nonconforming material. It was first developed for the production of material for the military but has been adopted for many civilian applications. The purposes of the process are to disposition the material (scrap, rework, or use-as-is) and to assure that corrective action is taken to prevent recurrence. Is this a failure cost because it arose out of a failure experience or a prevention cost because of its corrective action purpose? Consider the various audit activities. There are product audits conducted to provide assurance about the quality of the products being delivered. Process audits provide assurance that the processes are being operated in accordance with standards and are being maintained in a state of statistical control. Systems and procedures audits are designed to assure management that the processes they have put in place are in fact implemented, functioning as expected, and producing the desired results. How are these audits to be categorized? Are they even quality costs?

The objective of this discussion is not to inspire a feeling of hopelessness. On the contrary. In the area of quality costs the executive is not bound by someone else's system, designed to satisfy other objectives. In these waters every person is his own Columbus. A superbly useful management tool emerges when all concerned parties think their way through the issue of what information should be collected and how it should be organized to provide relevant insight into the processes of the company.

Just as the cornea of the eye provides the only noninvasive opportunity the physician has to look directly into the circulatory system of the patient, the properly designed quality cost system provides a superb and noninvasive opportunity to observe directly the most intimate processes of the business.

There are only three questions that matter. Is the motivational effect consistent with the behavior that management desires to stimulate? What cost is reasonable to bear in the interest of precision and elegance? And does the measurement obscure the objective? This simply demonstrates that we are not dealing with cost accounting questions. This is not an absolute science. Economic models are always controversial. The driving question remains. What organizational behavior are you trying to motivate?

JOURNAL: I've read all the material my quality manager gave me on the subject of quality costs. Not a whole lot of help, I admit. Called in the accountants too. Thought they could help. The bottom line was that they prepare the reports they were asked to prepare. They don't really understand them and obviously aren't very interested. Their lips curl a little around the edges when I speak of the quality cost reports as though they were products of the cost accounting group. I get a quick disclaimer. I infer they don't think that they're very respectable.

In reading all the stuff I was given, I found myself disagreeing with a lot of it: what to include, how to categorize it, etc. Kept wanting to get my fingers into it and rearrange it for them. Why?

*Maybe it's too product oriented. Of course it includes some of the costs that surround the processes that produce the product. But as I thought critically about what I was reading, I once again had that growing feeling of nausea that I experienced on New Year's Day. What they were missing was what I had been missing. They seem to go to excruciating lengths to collect all the nits and lice, but the failure costs! I began to realize that most of the problems we seem to have in the shop are really the by-products of quality problems—overtime, broken runs, excess capacity, idle capacity too. Something even dawned on me about inventory. We operate on the assumption that excess raw and work-in-process inventory is made up of parts that we have. That's dumb! Excess inventory is made up of parts that we **don't** have!*

I called in the material manager. Asked him to bring in his work-in-process inventory report. Scared the devil out of him I think. What's everybody so jumpy about around here? Come to think of it, I'm asking questions in some areas that I never have before. People are wondering what I'm up to I guess. Fair enough.

Asked him to pick out a few big ticket excess inventory items for us to look at together. He was sweating. He also had the answers. I think he knew more than he thought he knew. Turns out we have about 200,000 piece parts stranded in work-in-progress that we can't assemble into finished goods. I thought maybe our customer had canceled an order. My material manager says no. On the contrary, we're behind schedule with the customer and in big trouble. We'll wind up paying overtime, premium freight, the whole bit. It seems that we can't complete the assemblies because of one lousy part number. It's got a five-week flow time through the fabrication shop, and the whole order was scrapped at the last operation for a critical defect that had been built into the part at the third operation four weeks ago. That's when I decided that excess inventory, in these cases, is made up of the parts we don't have. Hundred-piece assemblies stranded in work-in-progress because of one lousy part!

*Called in the quality manager and asked him where **this** appeared in his quality cost report. You know what the answer was. Decided then and there that I **am** going to get my fingers into that report. From now on it's going to tell me what I want to know. We're going to redesign that thing to be something more than wallpaper. The quality manager got wide-eyed when I told him that. He seemed to get sort of happy/scared. Happy because I'm interested at long last and scared because he thinks I'm going to mess up his system. Afraid we won't do it like the book. To heck with the book. It's our business and our system.*

Also got to rethink how we report those costs. Is quality cost as a percentage of sales the right number? I'm not sure. What's the return on my investment in quality? Should I be investing more? Less? Hmm. Interesting. I just said investing. I'm beginning to think differently about this. We may be looking at an investment opportunity, not just a cost. This could get to be fun. And useful.

Choosing an Index

Choosing a basis for measuring quality costs is not a trivial matter. It is taken entirely too much for granted. Quality costs are usually indexed against some base for comparison purposes. If not the comparison to some other operation (which should be done with extreme care if at all), at least to prior periods, to discern improvement or deterioration.

The purpose of indexing quality costs is to remove the effect of some confusing, but irrelevant factor from the movement of the index. In other words, what affects the absolute level of quality costs other than quality itself? Next, how can we remove that effect so that it doesn't confuse us? Consider, for example, that the accounting calendar is such that there are months of different lengths. There are four-week months and five-week months, and of course, several of each have paid holidays in them. The quality costs will vary from month to month if for no other reason than the length of the month. We could become unduly alarmed or encouraged as we watched the number increase and decrease from month to month, when actually, what we would be becoming is unduly confused.

To remove the source of this confusion we might decide to divide the quality costs each month by the number of days. Our index would become quality costs per day. We have eliminated the effect of a varying number of days from the response of the quality cost index. In choosing a base, care must be taken to select the index that suppresses the proper effect. Otherwise one could find oneself taking up residence in a fool's paradise.

Sales

Normalizing quality costs against sales is probably the most common approach to indexing quality costs. It is also probably the most misleading.

But before exploring that, first consider the highly efficient fast-food franchise of your choice. Raw material (forgiven the pun) comes in during the early hours of the morning. Most of it has been consumed by closing. Finished goods are made to order or nearly so. None are carried over to the next day. Quality costs consist of scrap (product destroyed in process or discarded because it was not sold within a defined period after being prepared). After all, it is necessary to anticipate the noon rush by stocking up a bit before it begins. Pancake batter is mixed by the batch and not by the pancake. And to complicate matters, the pattern of buying as well as the volume is not the same on weekends as it is on weekdays. Also, Saturdays are not quite like Sundays, holidays are not like either, and winter buying patterns are not like summer patterns. Perhaps you let the employees take home any leftovers at the end of the day, so you might be a little concerned about abuse of this policy. In this case sales might be a pretty useful indexing tool. Quality costs per thousand dollars of sales could tell us a lot.

What made a sales base work in this case? One big thing is that scrap is contemporaneous with sales. Most scrap occurs on the same shift, surely the same day the product was made.

Unfortunately, this is not the case with most manufacturing operations. As the period of manufacture for a unit of product gets longer and longer, then the period in which the sale occurs gets further and further out of step with the period(s) in which the product was manufactured and the quality cost incurred. It might take a long time to build a boat. Most of the scrap and rework occurs during the fabrication and assembly period. We finish it on the last day of the month (it's amazing how much product gets finished on the last day of the month). The quality costs per thousand dollars of sales are enormous during the build period. There are no sales. But the following month we recover in a hurry. Scrap and rework are zero. Sales are big.

As the time the product spends in finished goods inventory grows longer, quality costs, which are incurred during manufacture, have less and less meaning when compared to sales. Warranty costs, which may extend for a long period following the sale, also need a different treatment.

The profit margin effect needs to be considered. If profit margins are modest and stable, then the margin effect is minor. But if margins are large, and particularly if they are unstable, then sales-based indices are of questionable utility. Quality costs drop to apparently modest levels if the profit margins are high enough. Why should higher margins justify higher quality costs? In this case we are introducing an inappropriate adjustment by using this index. And what about the case where margins are highly variable? Sales are held to move products that are excess in inventory or when seasonal goods must be cleared out. Influential customers, one way or the other, seem to enjoy better pricing. Finally, the business hires a world-class marketing person. This person opens new markets for the product and finds unique applications for which there are no competing products. Pricing is reviewed and profit margin doubled! Quality costs just fell dramatically. Or did they? All these factors and others can affect the dollar sales volume completely independent of the legitimate factors that affect quality costs.

Cost of Goods Manufactured

This is also a popular base and a useful one in many circumstances. But as always, one has to be careful. This index avoids the profit margin syndrome and is apt to be much more contemporaneous with the period of manufacture than sales in most manufacturing activities. But cost of goods manufactured is often recorded when products are completed and sent to finished goods. If the period of manufacture is relatively long, we may have a dislocation, with quality costs occurring in one period and the cost of goods manufactured incremented in another. Unless volume is such that it all tends to even out, one can be deceived.

More important, it gets tricky when you think about the consequences of using actual costs. Any excessive quality costs that we are experiencing will go into both the numerator and the denominator. While this does not erase the effect of increasing quality costs, it certainly suppresses it. More insidious is that as quality deteriorates, other performance factors tend to deteriorate with it. Labor efficiency tends to decrease. Setup costs skyrocket as production runs are broken into to make room for special orders to replace defective products. Overtime tends to increase in manufacturing. The list is endless. All of these are secondary effects of quality problems that go largely unappreciated. Quality is not the effect of these inefficiencies, it is their cause! Yet all these costs add to the denominator of our index and are probably not included in the numerator. Therefore, they actually suppress the response of the quality index! If we consider that inflation of the denominator "earns" an increase in the numerator, then all these inflated costs, the consequences of poor quality, "earn" or allow for more quality costs in the numerator. One may be suppressing the movement of the index by using actual manufacturing costs.

Standard Cost of Goods Manufactured

This is the author's favorite in most manufacturing situations! It is a very useful index. It avoids the problems associated with the use of a sales base as well as those associated with the cost of sales problems discussed above. As variances to standard costs accrue, they do not inflate the denominator. This allows the increasing quality costs appearing in the numerator to have their full effect on the index.

Process Hours

In process-driven situations quality cost per process hour is a useful index. It is insensitive to the number of shifts worked and whether or not they drew premium pay for the operators, which are both disadvantages of the use of an actual cost base. Its use is indicated in situations when the number of operators is dictated by the process and process quality tends *not* to be labor dominated, but machine or material dominated. In these situations the accounting function, manufacturing engineering, and production tend to be process centered, and many other costs may be evaluated on a process hour basis. It is always good, when practical, to manage quality costs in the same terms that other cost elements are managed. This avoids the problem of asking others to become "bilingual," having to speak one language when dealing with quality costs and another the rest of the time.

CHAPTER 9

Reporting Quality Costs

JOURNAL: I'm glad I got into those quality costs. One thing is certain. I'll never again ask my quality guy how our quality costs compare to those of our competition. I'm sure no one else in the entire world collects them as we do. That may be an edge for us. I think we will have a much more useful insight into the impact of quality on the productivity of our operations than our competitors have.

*As I got into restructuring our quality costs, I developed an itch that I had a hard time scratching. Was I prostituting the quality cost system and turning it into a productivity report? I decided not. First, quality has a lot to do with productivity. I'm having an increasingly difficult time separating them in my mind. The quality cost report is a productivity report. Let's face it, our customers couldn't care less how much it costs us to produce our products or how much scrap and rework we incur in the process. All they care about is how much our products cost them and how good they are when they get them. The quality cost reports deal in production economics. They **are** productivity reports. They measure waste and they test the effectiveness of the discretionary activities we engage in.*

With that thought behind me, I made some heroic changes, at least to hear my quality guy tell it. First, I changed the prevention category. I threw out all the bits and pieces of time that people spend assuring themselves that they have done their own job properly. I suspect we saved a lot of money there, just in the cost of data collection. Prevention will be the preventive activities of the quality department only. Period! I know it's a compromise. I think I can handle that. But I almost went nuts when I found out that the quality director, his secretary, and the whole administrative function was categorized as preventive. That's crazy. After I made that adjustment, I realized just how little of the quality department's effort was really preventive. The costs I removed are necessary quality costs to be sure. They've got to go somewhere. So we created an administrative category. Makes sense to me. By this time my quality guy was beginning to get the hang of it although he was a bit nervous about having his office's costs displayed for all to see like that. Then he said that he wanted to make sure that the time spent by his quality engineers mollifying upset customers was logged as a failure cost. He feels that true corrective action, which we both agree we want categorized as prevention, is too easily mixed up with noncorrective activities incidental to a problem. He also feels his quality engineers aren't being used strategically. Good point. The airline tickets and the three-martini lunches are also failure costs.

We agreed to treat appraisal costs a little differently too. That category will include institutionalized inspection regardless of the department in which it occurs. That means we will include all QC inspection. We will also include those dedicated "sorters" that manufacturing uses to screen the product before it goes to inspection for acceptance. All they do is stand at the end of the line and inspect. Including them might motivate manufacturing to improve their processes and get rid of that cost. Now that I think of it, I think manufacturing has those people included in their standards so that they don't even show up as a labor variance!

Internal failures will include scrap and rework, as they have in the past. Manufacturing almost flipped when I told them it would also include process scrap. They consider that normal production scrap, borings and shavings kind of stuff. Well, it is normal. It's also scrap. We have a lot of process-intensive departments. Flash, edge trim, and so forth is easy to live with if it's in the standard and you don't have to answer for it. No pressure for process improvement. Including it may help to force attention to process quality. I did agree to keep it separate from the nonconforming material kind of scrap, however. Don't want to get confused on that point. Manufacturing has been pushing the quality department to report only variance from standard. Convenient, since they include an allowance for scrap and rework in all of their standards. Quality has hung tough on that one, thank goodness. We have been reporting and will continue to report all failure costs, not just those above the standard allowances.

Haven't figured out the external failures yet. I want to get everything I can identify into that category. Those are the most critical quality costs we have. I don't want any to get away. I've also asked our accountants to see if there isn't a practical way to include all the secondary effect of quality problems, internal and external. It may not be practical, but at this point I'm convinced more internal failure costs are being left out than are included.

Now, how can they be reported in the most meaningful way? The tab runs and the pie charts leave me a little cold. Actually, they're what I asked for, but looking at these costs in the new light, I'm not sure that they tell me what I need to know. Maybe we ought to turn the statisticians loose on that one. Control charts in administrative areas? Never thought of that. Why not?

Quality Cost Reports

Quality cost reports take many forms. At one extreme is the technique of periodically depositing massive tab runs on the desks of managers that contain all the detail sorted by work center, department, part number, cost category, and so forth. While impressive in bulk, these reports usually serve only one useful purpose—reading them is a sure cure for insomnia.

Then there is the graphic approach. This one management often inflicts on themselves. "Just give me a few graphs so that I can quickly capture the essence of the situation." Although this sounds plausible on the surface, there are a few secondary questions to consider. Probably the executive doesn't ask his controller for a few charts and graphs regarding profits, margins, manufacturing costs, and so forth so that he can "capture the essence of the situation." He may have charts *also*, but definitely not *instead*. Charts are a useful supplement to detail, but not a substitute. Both are needed. We should, however, avoid the confusion of data with information. The detailed tab runs are usually more of the data sort. Data arranged and organized effectively produces information. Information is what is needed to make decisions.

Clearly, one useful statistic is quality costs as a percentage of a selected base. It is useful to be able to say that quality costs are running "xx% of standard cost of production" for instance. Such a statistic should probably be prominent in every reporting system. It lends itself to line graphing. Statistical control charts using quality cost percentages are very useful.

Cost Avoidance

Insight is often gained by tracking quality cost avoidance. This certainly fits into management's way of thinking about their businesses. If management has been persuaded to consciously invest in quality improvement, then they are entitled to a return on that investment. One form their investment takes is cost avoidance. Cost avoidance in this instance must be volume adjusted. Specifically, quality cost avoidance is a volume-adjusted cost reduction from a base period. The question is this: If the business was spending quality cost dollars at the same rate this period as during the base period, what would the quality costs be? The difference between that number and the actual quality costs is quality cost avoidance. It can be expressed this way:

> (Current volume of the denominator) × (base
> period percentage/100) − (current period
> actual) = quality cost avoidance

Obviously, cost avoidance can be positive (an improvement) or negative (a deterioration).

Mix Effect

Assume that quality costs are being tracked at the department level, the plant level, and the corporate level. A curious situation surfaces when one calculates cost avoidance that is not so obvious when dealing with percentages. It's there with percentages too, you just don't see it. The whole almost never equals the sum of the parts! The cause of this phenomenon is mix effect. One would expect that the sum of all the departmental cost avoidances would equal the total cost avoidance calculated at the plant level. This will rarely be the case.

Each department has a different quality cost percentage factor. Some will probably be quite different. As long as the percentage of total activity represented by each department remains the same between the base period and the period being reported, then the sum of the departmental cost avoidances will in fact equal the total. Rarely, if ever, will this happen. Consider Table 9.1. Quality costs in department A were $18,200 in period 1. This amounted to 7.0% of the standard manufacturing cost of $260,000. If department A had still been operating at 7.0% of standard in period 2, their quality cost would have been 0.07 × $250,000, or $17,500. Their costs were actually $17,000. Therefore, they enjoyed volume-adjusted cost avoidance of $500. This is the amount shown in Table 9.2.

Table 9.1   *Quality Costs by Department*

Department	Period 1			Period 2		
	$	SMC	%	$	SMC	%
A	18.2	260	7.0	17.0	250	6.8
B	16.8	280	6.0	14.8	260	5.7
C	11.0	270	5.0	9.7	220	4.4
D	19.0	190	10.0	26.1	270	9.7
Total	$65.0	$1000	6.5%	$67.6	$1000	6.76%
Cost avoidance				($2.6)		

SMC = standard manufacturing costs. All dollars in thousands. For convenience, total standard manufacturing costs have been set at one million dollars in each period.

Table 9.2   *Cost Avoidance*

Department	Cost Avoidance
A	$ 500
B	800
C	1300
D	900
Total	$3500

All dollars in thousands.

   Total actual costs in period 1 are $65,000. This is 6.5% of the total standard manufacturing cost of $1,000,000. The second-period standard cost was also $1,000,000. Had they still been running at a rate of 6.5%, the costs would again have been $65,000. However, the actual costs were $67,600. This is a negative cost avoidance (cost increase) of $2600. But the sum of the departmental cost avoidances is a positive $3500. Somehow $6100 of cost avoidance has disappeared!

   On careful inspection we find that departments A, B, and C declined in volume. Their quality cost factors ranged from 5.0% to 7.0% in period 1. This decline in volume was replaced by an increase in department D. Department D, however, has a quality cost factor of 10.0%. We replaced a lot of 5% to 7% work with 10% work. Department D improved in period 2 from 10.0% to 9.7%, but this rate was still well above that of the volume they replaced. This is mix effect. While we are used to dealing with the product mix idea, the same phenomenon is often overlooked when we are dealing with departmental mix or plant mix. *Any time one is dealing with a*

QUALITY COST ANALYSIS

(\$000 OMITTED — % OF STD COST OF GOODS MANUFACTURED)

TOTAL	FY 87 \$	%	SEP 88 \$	%	OCT 88 \$	%	NOV 88 \$	%	FY 88 YTD \$	%	YTD COST AVOID
STD COST GOODS	227,124		22997		24454		27655		259,686		
QUAL DPT	1,500	.7	156	.7	138	.6	148	.5	1,759	.6	44-
INSPECT	4,782	2.1	434	1.9	463	1.9	441	1.6	5,010	1.9	458
INT FAIL	8,433	3.7	883	3.8	1081	4.4	1216	4.4	8,459	3.2	1,183
EXT FAIL	995	.4	86	.4	88	.4	138	.5	1,025	.3	113
PROC SCR	7,396	3.3	622	2.7	731	3.0	874	3.2	7,723	2.9	733
TOTAL	23,106	10.2	2181	9.5	2501	10.2	2817	10.2	23,975	9.2	2,444
COST AVOIDANCE OPERATIONS			166		16		23				2,688
MIX EFFECT			8-		29-		26-				245-
NET AVOIDANCE			159		14-		4-				2,444

Figure 9.1 *Quality cost report.*

118

consolidated report, that is, a report that consolidates a number of quality cost–generating entities, mix effect is present and must be isolated if the analysis is to have any meaning. Specifically, this means that corporate quality cost reports that consolidate businesses or factories and factory reports that are made up of departments are fundamentally flawed if they do not address mix effect!

Types of Quality Cost Avoidance

There are two kinds of quality cost avoidances. One type, which we can call operating cost avoidance, is the sum of the cost avoidances of the operating units. In the case of our example the units are departments. The other important type, which can be called net cost avoidance for want of a better term, is the cost avoidance calculated at the aggregate level. This is the amount that actually goes into the cash register. The difference between these two is mix effect. A report reflecting these might look like that shown in Figure 9.1.

Figure 9.1 is a sample quality cost display designed to be available on computer in the offices of concerned managers in a major manufacturing company. It reports the plant level costs with consolidations at the business unit and corporate levels. It displays the current quarter by month and has the ability to scroll forward and backward in time. Cost avoidance is calculated by quality cost element as well as in total by month. Monthly and year-to-date cost avoidances are separated into operating cost avoidance, mix effect, and the total or net cost avoidance. At the business unit level, consolidations are available that show the distributions by plants that make up the total, in addition to the summary by cost element shown here. At the corporate level, summaries are available by business unit as well as by cost element. Mix effect is arrived at in this system by calculating cost avoidances at the plant level and at the consolidated level. The difference between the sum of the plant cost avoidances and the amount calculated at the consolidated level is mix effect.

Statistical control charts (Figure 9.2) and bar charts of cost avoidance (Figure 9.3) are also available in the system. This is a very practical, as well as paperless, system. A manager in the corporate office can look at his quality cost summary by category and tell in an instant what his costs are, what elements are generating them and how much of the result is due to operations and how much is due to mix. If so moved, he can call up a report that shows cost avoidances by business unit to determine the locus of any difficulties. Having isolated the offending business, if that is the nature of his quest, he can look within the business to isolate the offending

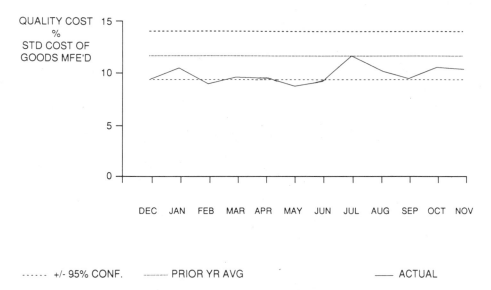

Figure 9.2   *Control chart.*

plant. He can then display the plant and identify the responsible element or elements.

The control chart is also very handy. There is a strong tendency when the news is good for executives to attribute the performance to their personal brilliance, discounting out of hand the possibility of blind, dumb luck. If the news is bad, on the other hand, it is usually attributed to some factor clearly beyond their control. The control chart works as well in offices as it does in factories. It aids in separating variation that is normal to the system from the effects of assignable or nonsystematic variation. Much valuable time is wasted in executive conferences speculating over the source of some trivial improvement or deterioration when the truth is they are looking at normal month-to-month variation. Such are the ways of the potentates.

Mix of Quality Cost Elements

In analyzing quality costs the mix of cost categories often offers more useful insight than the absolute numbers or the level of the costs when compared to some base. When quality costs are looked at for the first time in a

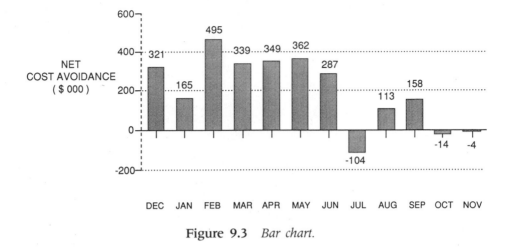

Figure 9.3   *Bar chart.*

production operation, it is not uncommon to find that consequential costs are running 75% or more of the total. This raises an obvious question. Is it not conceivable that an intelligent investment in one of the discretionary categories could produce a larger than offsetting reduction in consequential costs? We may have a very lucrative investment opportunity before us.

Appraisal costs are usually the largest part of the discretionary segment, often comprising twenty or more of the twenty-five points of discretionary costs that we might be looking at. If, in spite of this investment, internal failure costs are modest and external failure costs are high, we should be concerned about the effectiveness of the inspection activity. Perhaps the failing characteristics are not inspected or perhaps they are not even inspectable. If functional failures after delivery are the big contributor to the external failure costs, perhaps a reliability improvement effort would pay more dividends. If internal failure costs are high and not improving and the inspection effort appears to be effective and the preventive category appears adequate, we might look at the way the preventive dollars are being spent.

In studying these relationships we would expect to find certain characteristics in an effective and efficient operation. To generalize from an earlier chapter, efficiency has to do with doing things right. Effectiveness has more to do with doing the right things. As the effort matures, more and more of the discretionary activities should become internalized in the other functional activities of the company. Thus the discretionary costs would fall in absolute value, but since the "pie" is also shrinking, they would still loom

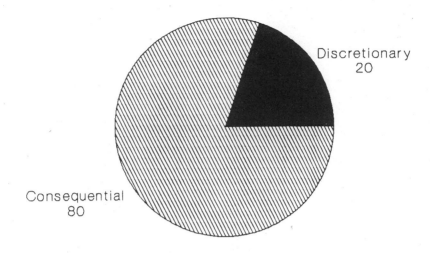

Figure 9.4   *Quality cost distribution by class.*

large as a percentage of the total. As preventive efforts take hold, both internal and external failure costs would decline. Presumably internal failure costs would decline significantly and external failures would virtually disappear.

In the final analysis we would expect to see a drastically reduced level of quality costs as a percentage of cost of production, for instance. Once in equilibrium, that amount might be about equally divided between discretionary and consequential costs. Prevention costs would represent a larger proportion of the discretionary category, the portion dictated largely by the nature of the product and the processes that produce it. The vast majority of the consequential costs would be internal failures. External failures should be a trivial amount of the total—trivial in amount, but massive in the attention they attract. External failure costs can never be taken for granted regardless of the level. Protecting the customer against the consequences of unsatisfactory product is a fundamental obligation of any product or service provider.

To reveal these relationships "pie charts" are popular. They must be used with care, however. Consider Figure 9.4. Discretionary costs are running 20% of the total. An excellent opportunity for improvement would appear to exist. Figure 9.5 shows distribution of those costs by category.

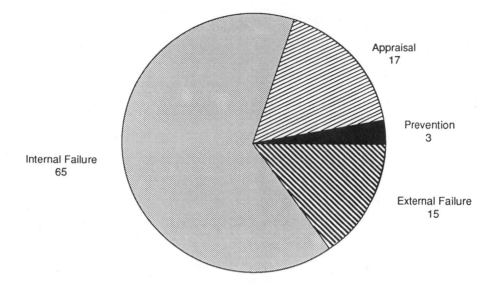

Figure 9.5 *Quality costs by category.*

Prevention costs are modest and perhaps overstated. If, on the other hand, they are as represented, then we should probably question the effectiveness of the preventive program in light of the consequential costs. Internal failure costs are massive and external failures are significant, at least in a distribution of percentages of the total.

Figure 9.6 shows the distribution after a concerted attack on the cost of quality. How do these figures compare with those shown in Figure 9.5? Preventive activity has certainly expanded. But so has appraisal. A success-ful preventive program should allow one to reduce appraisal activities over time. Perhaps that time hasn't arrived yet. In spite of the increased ap-praisal costs, internal failures remain high. Although this is not inconsistent with high appraisal costs, it does bring into question the effectiveness of the preventive activity. External failure costs have been reduced to a modest level but are never to be ignored. So what does all of this tell us? Well, not as much as it should, actually.

Both charts deal with the distribution of 100% of the quality costs. This is a fixed amount. But quality costs aren't fixed! The pie chart will always total 100% regardless of the absolute amount of the quality costs. Had failure costs been totally eliminated and appraisal costs minimized consis-

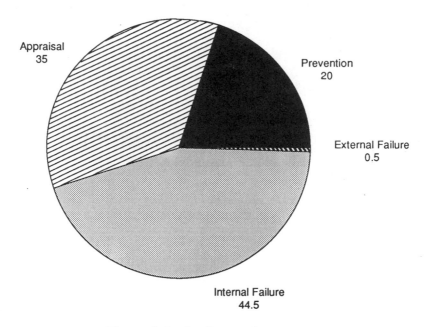

Figure 9.6   *Quality costs by category.*

tent with the processes being governed, we would find that the discretionary cost elements were massive as a percent of the total (100%) and we could easily become disoriented.

It is recommended that when pie charts are used they should be dollar based, not percentaged based, and when comparisons are made, proportional area pie charts should be employed, as in Figure 9.7. Pie chart

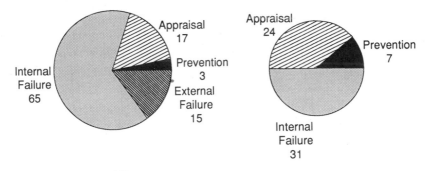

Figure 9.7   *Proportional pie charts.*

comparisons using percentage distributions should be used with extreme care. In Figure 9.7 we find that in the original situation quality costs were $100,000, in addition to being distributed as in Figure 9.5. We also learn that they have been reduced to $62,000, a 38% reduction. Prevention costs have roughly doubled from $3000 to $7000, not the magnitude implied by Figures 9.5 and 9.6. Appraisal costs have in fact increased. External failure costs have disappeared when we round to the nearest thousand dollars and internal failures have declined over 50%. Most would agree that this is a more useful way to look at changing distributions.

CHAPTER 10

Analyzing Quality Costs

*JOURNAL: We just had our first victim. It was bound to happen sooner or later, I guess. One of our senior guys. Just couldn't go along with the program. It stuck out all over him. His body language was incandescently clear. This whole crusade was an amusing joke being put on for his personal entertainment. Never said a word in opposition, just managed to ignore the whole thing. When forced to acknowledge our new directions within his organization, he managed to do it in a way that ridiculed them at the same time. He was a very good functional performer. So was his organization. As a result he thought he was bulletproof.. I spoke to him repeatedly about his attitude. All I got was a wide-eyed "Who, **me?"** Had his whole organization stopped cold in its tracks. I'm sure others were watching to see what I'd do. A litmus paper test of sorts for the "old man." So I fired him! Did it myself. A leader has to do his own dirty work, clean up his own messes. Usually human resources gets to do the messy part. Not this time. Replaced him with a guy who isn't his functional equal but believes in what we're trying to do. Made sure we back-filled with the same kind of people down the line. Was pleased to find out that even the kid got a promotion out of it. Now that he's in* *management himself it's going to be interesting to watch. Seemed to take about thirty minutes for the word to circulate through the organization. Has made a difference already. Yes, folks. I really mean it. And I'm not just some harmless old goat with a new toy. Leaders need teeth too.*

This quality cost stuff is getting interesting. Got some useful reports flowing. Accounting, with help from others, was able to recover some history too. Now we don't have to wait forever for trend data to develop. The problem now is that the analysis is getting complicated. These aren't all contemporary costs. Neither do all of the cost elements behave the same way. I'm concerned that a lot of bad decisions can be made while people are learning to use the new tool. Do-it-yourself training with hand grenades is a tough way to go. Decided to kill several birds with one stone. I've asked that all our operations reviews start with a comprehensive report on their quality costs together with their analysis of the data and planned actions. This tells people that I'm serious, it assures me that they're looking at the data and analyzing it, and also gives me some insight into the quality of their analysis. Must also admit that I've learned a few new tricks. Some of these people have come up with some clever insights. I've then been the agent transferring these into other organizations. Feel like a bumble bee pollinating flowers, but what the heck.

Operating Characteristics

The different categories of quality costs have different operating characteristics, and this fact is of tremendous tactical and strategic importance in the management of the quality program. Consider the relationship between categories and the time constants involved in their application.

Prevention

A classic prevention activity is quality engineering effort to improve designs and processes. Prevention activities vary inversely with failure costs. By definition, any problem prevented can never manifest itself as scrap, rework, or warranty action.

On the other hand, it takes time for the quality engineering effort to show results. Data must be collected. Processes must be studied, as must designs, tools, work instructions, and test equipment. Changes must be initiated. These must work their way through the process to implementation. All of this takes time. And the amount of time is also a variable. Early results may come relatively quickly but probably not immediately. Obvious corrections that are totally within the control of the manufacturing department may be made in a day or two. As these more obvious changes are made, the study and correction period draws out.

Preventive activities are very cost effective. A process improved continues to pay dividends long after the effort required to effect the improvement is terminated. Things fixed tend to stay fixed without further preventive effort. This is why prevention programs can be effective and at the same time represent a small fraction of the total quality cost.

Preventive activities are unusually sensitive to the strategy with which they are employed. It is sometimes assumed that if we do preventive things, prevention results will automatically follow. Nothing could be further from the truth. The intelligence of the strategy with which preventive effort is deployed is absolute critical to its success. Preventive strategies are definitely *not* "no brainers"!

Appraisal

What, then, are the operating characteristics of inspection? First, inspection has an immediate effect. It is responsive. Insert inspection and it begins to show an effect immediately; products begin to be rejected. Second, any process improvement effect is strictly incidental to the inspection process. Third, it is expensive. Since inspection alone does not improve the process, defective products continue to be made and inspection must be continued forever. Finally, it is not particularly efficient. Good parts get rejected with the bad, bad parts escape, and the effectiveness of inspection is heavily influenced by the quality level of the material submitted to it. Inspection prevents escape. Period!

Administrative Costs

Administrative costs tend to be fixed over a broad range of production volume, whereas inspection costs tend to be variable with volume and quality engineering costs are somewhere in between. Administrative activities, depending on how they are defined, also rarely make any direct contribution to either prevention or appraisal. This alone is reason enough for wanting to separate them from the more productive efforts for monitoring purposes. Administrative costs could include, in addition to the quality head and his secretary, procedures people, budgetary analysts, and any "gofers" that seem to accumulate at the upper reaches of many large organizations. These are necessary costs of doing business, but we mustn't allow ourselves to become confused by their treatment in the quality cost system.

Internal Failure Costs

Internal failure cost tend to be contemporary with inspection activity. They also tend to rise and fall together. This can be deceptive. Eliminate inspectors in a department and watch quality costs fall. The classic lament of the unenlightened is "We didn't have all these scrap and rework costs until we hired those inspectors!"

Also, internal failure costs tend to be contemporary. The parts are made, inspected, and scrapped or reworked in a fairly confined period of time.

External Failure Costs

External failure costs are, of course, the most devastating of all. They wound the customer, and that is not what one is in business to do.

External failures tend to be dislocated in time. It may be weeks, months, or even years before the true impact of external failures is felt. In the meantime, the pipeline can be filled with products doomed to poison the marketplace.

Process Scrap

Process scrap tends to be fixed to the volume of production. It can be predicted within narrow limits and therefore is comfortable to live with. Therein may lie its greatest threat. It can't be attacked with inspection. The only attack on process scrap is a discretionary process improvement project.

A Scenario

Consider that a fabricating department is producing an unacceptable level of defective parts, which are forwarded to an assembly department for use. The parts take two weeks to go from the back door of the fabricating department to the front of the assembly line. This is a regrettable condition. It is also common.

Once in the assembly department, some of the parts must be scrapped. Others are reworked, for they are needed to satisfy production requirements. Others just go together awkwardly giving rise to excessive assembly time and degraded function and reliability once in the hands of the ultimate consumer. The full effect of the function and reliability problems isn't fully known because the pipeline to the ultimate consumer is long and the expected life in the hands of the consumer is also long. Should the product fail a year or so after being put in service by the customer, it will be a serious problem to the manufacturer whether the product is warranted or not.

An enlightened management immediately institutes 100% inspection to assure the quality of the product going to assembly. As this is not sufficient in and of itself, several quality engineers are assigned to the fabricating department to find the cause of the problems and institute corrective action.

The inspection has an immediate effect, and parts begin to be rejected in the fabricating department at a record level. Failure costs go up. Schedules fall into jeopardy as batches of material get rejected. Overtime follows as parts have to be reworked or replaced. The good parts now trickling out of the fabricating department begin their journey to the assembly area, moving with a bit more urgency now because the pipeline is drying up.

The screams of the assembly department continue. They still have to work their way through their backlog of stock. Returning it to the fabricating department would do little good since their rework processes are glutted and their capacity is completely consumed producing makeup parts. Besides, the assembly department desperately needs the parts in order to meet their schedules. In short, costs have gone up and organizational trauma has increased. Things have gotten worse in the fabricating department and are no better in the assembly department. If quality in the fabricating department is sufficiently bad, the assembly department will sooner or later find that they have been traded a supply problem for a quality problem. There just aren't enough good parts surviving the inspection process to meet their needs.

So far the return on all this investment and effort has been modest. Inspection has been added and will have to be continued since basically it is a sorting operation. Parts are still being scrapped and reworked, at heroic levels. At best, we are transferring the failure costs from the assembly department to the fabrication department, where they properly belong.

It didn't take long for this state to develop. After all, inspection has its effect immediately. A house of cards doesn't crumble in slow motion. In the meantime, the quality engineers have not been sitting on their hands. Thanks to the increased inspection effort, the quality engineers have some objective data available to them that did not exist previously. Data can be a valuable by-product of this type of inspection if the operations have the presence of mind to collect it.

Their first action was to assess the capability and the state of control of the offending processes. A word about process capability and control is in order here.

By "process capability" we mean the ability of the process to consistently produce products that vary *within* the limits allowed, the part tolerances in other words. The output of most industrial processes produces the classical "bell shaped" distribution. If that process is capable, then the distribution of individuals of the part parameter represented by the curve is totally contained within the allowed tolerance. Clearly, this concept is sensitive to both process centering and process variability. If the variability is marginally acceptable, as in Figure 10.1a, then the process must be kept precisely centered to avoid creating defective products. If the process output fits comfortably within the allowed limits, as in Figure 10.1b, then the quality of the output is not quite so sensitive to centering. In short, it becomes more difficult to manufacture a nonconforming product. Consider, however, a vastly incapable process such as the one we are probably dealing with in this scenario. It might look like Figure 10.1c. The output of the process exceeds the limits of the specifications. Inspection is used to cull out the defectives, those items in the tails of the distribution beyond the tolerance limits. Make the convenient, if absurd, assumption that the inspection is 100% effective. The output of the department following inspection is as shown in Figure 10.1d. It is intuitively obvious that the output of process (b) will assemble more readily, produce less variability in the assemblies, and in most cases produce a more functionally reliable product. Variability is a quality issue! A less variable product is a higher quality product in every way.

By "control" we do *not* mean a process that is doing what we want it to do. "Control" as used here is an abbreviation for "statistical control." Avoiding the technical definition of that term, which might not help too much at this time anyway, let's just say that a process is in statistical control when it is behaving predictably within defined limits of variation. In other words, there is no evidence of chaotic instability that would prevent us from predicting its performance.

To make these determinations the quality engineers begin collecting data with which to analyze process performance. As we might expect in a case like this, they determine that the process is *not* in a state of statistical control. In fact, the critical product parameters are bouncing around wildly showing the influence of apparently unrecognized process variables. The

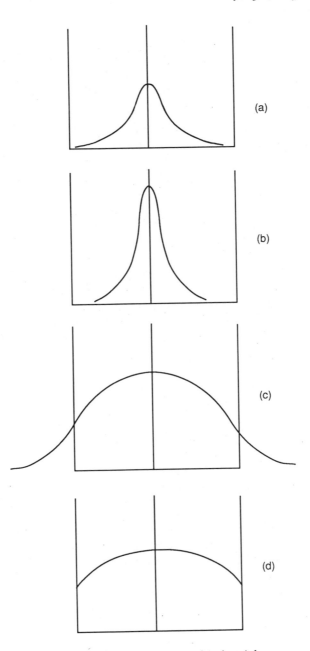

Figure 10.1   *Distribution curves of industrial processes.*

first order of business is to establish a state of process control. It is impossible to evaluate, let alone improve, a process behaving chaotically. Attempting to do so will only make matters worse. This is frustrating news to anguished managers and executives who are used to having their every command instantly obeyed.

Eventually, and this takes time, a state of statistical control is established. When managers look at process capability, they find, as is often the case, the process is operating off center and its natural variability is substantially wider than the part tolerance will allow. They've only learned what they already know, to the growing consternation of management. The process is making a huge amount of nonconforming material.

Centering the process is relatively easy and this makes a big improvement in the amount of defective production being made. Scrap and rework go down, but 100% inspection must continue because the process is still varying beyond specification limits. Now the hard part: reducing process variability. The quality engineers determine that some tooling is in bad shape and must be replaced. This takes time. One by one, they uncover other sources of excessive variability and take action to eliminate them. All these corrections take time. But one by one, as they are implemented, the process variability improves, until at last the capability is comfortably within tolerance requirements.

Let's reprise. As inspection was applied, the first effect was for internal failure costs to increase in the fabricating department. Depending on the length of the pipeline and the backlog of supply in the assembly department, the time came when failure costs in the assembly department began to fall. But equilibrium was reached rather quickly. Since processes had not improved, they continued to yield defective products. Scrap and rework remained high in the fabricating department.

As the quality engineering efforts took hold and process capability began to improve, the scrap and rework bill continued to decline. Moreover, the assembly department began seeing a less variable stream of product coming to them. Uniformity of supply increases productivity in a number of ways, some obvious, some subtle.

As processes became capable, 100% inspection was no longer required. Statistical control charts were instituted and the manufacturing operators were trained in their use. After a reasonable coaching period, the quality engineers were redeployed to other situations.

Life at this point is delightfully dull. Few will miss the "good old days."

What has been happening to external failure costs during this time? In this case, since the suffering party was the assembly department, a surrogate for the external customer, little of benefit to the ultimate customer occurs at first. The assembly department has been culling out the defectives and what has gone to the outside customer has been nominally conforming product. All that rework and variability in the assemblies certainly did not help the uniformity of performance in the field. How deleterious that performance variability is we frankly don't know at this time. This variability in assembly could also affect product reliability and life expectancy. Again, we don't know; we hope.

If these external considerations are of such a nature as to result in unsatisfactory product being delivered into the stream of commerce, then the consequences will be profound. External failure costs will be high due to warranty claims, field service expenses, and customer returns—the full gamut of woes, their nature dictated by the nature of the product and whether or not we are selling to an industrial customer or the ultimate consumer. The time for the customer/consumer to see the fruits of the massive inspection program depends on the length of the pipeline and whether it was purged as the first step in the program of inspection triage.

All of this is summarized graphically in Figure 10.2. At T_0 (time zero) total quality costs are high and failure costs are unacceptable. At T_1 the level of inspection is increased, with the preventive level remaining the same. Internal failure costs increase instantly, with no accompanying decrease in external failure costs. At T_2 additional preventive effort is brought to bear. Neither internal nor external failure costs are affected. At T_3 internal failure costs begin to decline as the preventive effort takes hold. External failure costs are as yet unaffected and the inspection level, of necessity, remains high. At T_4 external failures (finally) begin to decline. Internal failures continue to decline but inspection remains high. Processes are not sufficiently capable to allow reduction in inspection. By T_5 external failures have been eliminated. Internal failures have been reduced and are in the area of diminishing marginal return for the type of preventive work we have been doing. Inspection has been reduced to an audit level and the preventive resources have been largely redeployed.

Dealing with Time Shifts

One of the most frustrating aspects of evaluating and managing quality costs is dealing effectively with the time shifts depicted in the operating characteristic curve. Think for a moment of a quality cost report as a "spread sheet" with months across the top and the quality cost elements down the side. The tendency is to think that all of the costs in each column

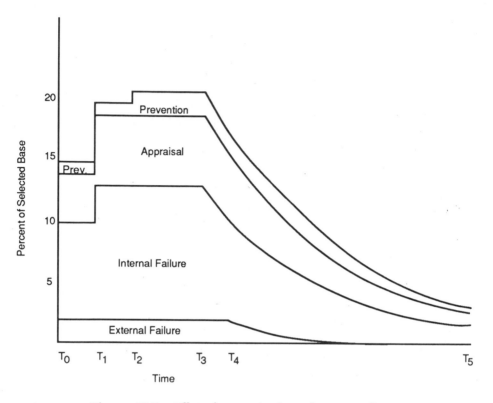

Figure 10.2 *Effect of preventive inspection on quality cost.*

go together. This is the accounting influence. Although they do, in the sense that all of July's costs were *recorded* in July, it is self-evident that all costs recorded in July were not rooted in July (the economist's point of view). If we conveniently assume a cause-and-effect relationship between the elements appearing in any given month, we may be guilty of overly simplistic thinking.

Prevention

We said earlier that there tends to be a delay between the application of a prevention activity and the realization of results in the form of reduced costs in other categories. Moreover, these cost reductions, when they do occur, tend to be cumulative. That is, once an improvement is institutionalized, it tends to continue to pay dividends month after month without further effort. This is why prevention programs can be highly effective while representing a small ongoing investment relative to other cost elements. As a consequence, the cumulative good attributable to prevention effort is difficult to assign to any given month. It is sometimes possible with microanalytic techniques to identify the effect of each preventive action and to evaluate their impact on succeeding months from a base period. But while often possible, it is rarely practical to do so.

Appraisal

Appraisal costs do tend to be contemporary. While accounting practices can distort the reporting, this is unusual and rarely has any practical effect. Inspection activities conducted in July, for instance, are recorded against July and relate to July's activities.

Administrative Costs

Administrative costs are straight period costs and typically show no interraction with other cost elements.

Internal Failure Costs

Internal failure costs—scrap and rework—generally tend to be reported in real time, that is, when they were incurred. However, there still may be some distortions. Rejected parts may be pushed aside in production and allowed to accumulate until a more convenient time to deal with them. Sooner or later they are sorted. A quantity of scrap may be generated at that time. However, the parts were manufactured and inspected in an earlier period. The replacement parts may be manufactured in a subsequent period. Highly cyclical production will invite this kind of treatment. During rush periods defective product is accumulated to be dealt with during slack periods.

It is not unheard of for manufacturing to deliberately "bank" defective product. A supervisor may decide that this month has been difficult enough as it is. A heavy bill for scrap and rework would be sure to invite a lot of hostile questions from others in the organization. Maybe next month will be better. If he should generate favorable variances in next month's operations, then he could absorb some of this rework without creating problems in the cost reporting. This thinking is not restricted to first-line supervision. Neither does it necessarily suggest the actions of those in the system trying to "get away with something." It may simply represent misguided thinking. Why should the company handle, warehouse, insure, and pay interest charges to store defective material until a later date when it could and should be disposed of now? This is clearly another hidden quality cost.

The net effect of banking scrap and rework can be quite exciting. It is not unusual in manufacturing operations for the last month of the fiscal year to contain some surprises, usually ugly ones. These are to a large degree the accumulated consequences of the twisted idea that one should never do today what can be put off until tomorrow, particularly if it is unpleasant. But at the end of the year, what with year-end closings, income taxes, annual reports, and auditor certifications, the time of reckoning arrives. Manufacturing has been banking rework and scrap, waiting for a more fortuitous time to deal with it. As is often the case, times don't get that fortuitous. In the meantime their bosses have been banking other writeoffs also waiting for a more fortuitous time. This policy works about as well for them as it does for their subordinates. Simultaneously, the accountants have been postponing the writeoff of some obsolete inventory, obsolete because it was put into finished goods and subsequently discovered to be defective and requiring scrap or rework.

All of this falls under the heading of "betting on the come." But what's coming seldom falls under the heading of good news.

External Failure Costs

External failure costs, which are the most exciting and usually, in the long run, the most damaging category, are in general the most severely distorted by the time factor. Upon completion of manufacture products go into the finished goods inventory, where they may remain for some time. This depends in part on the nature of the business—seasonality in sales, Christmas decorations, for instance—but also on the way in which we manage inventories. A first-in, first-out approach is recommended in most cases. However, first-in, never-out also seems to be a popular model. The product may go from finished goods to a mixing warehouse somewhere where customer orders are filled. In other cases large commercial products may go to a customer location and wait for installation or erection. Consider a newspaper printing press that often waits for a building to be wrapped around it before it is completed. Other products may go to commercial customers who also inventory them for a while before they are sold to the ultimate consumer. Products found to be defective upon being put into service may have been manufactured a long time ago! Add to this a warranty period, and a company can be incurring external failure costs for a very long time following the manufacture of the part. It has been wisely said that "all management mistakes wind up in inventory"—a profound, if disquieting observation.

A particularly disturbing adjustment to be made in some businesses has to do with reserves for warranty. In some businesses it is common to have substantial warranty claims filed by commercial customers. These claims may or may not be well founded. Investigation may take some time. The company is required by accounting standards to recognize the liability of these pending claims on the balance sheet. This is done by setting up a reserve to pay the amount of the claim. Upon completion of the investigation, the claim may be paid in greater or lesser amount than the original claim. At that time the reserve will be decreased by the amount that was originally set aside, and the actual amount of the paid claim will go into the claim account. The difference between the amount of the original reserve and the amount actually paid will either add to or subtract from current-period profit.

The claim account is included in the external failure costs. Under the typical quality cost system, the claim hits the quality costs at the end of the investigation when the claim is actually paid. This makes sense on the surface but fails to stimulate the proper management response. When management sees a large increase in the external failure costs one month, they will probably inquire as to the reason. When told about the claim that was booked, their response is apt to be "Oh, we reserved for that claim six months ago," meaning that the claim hit the balance sheet six months ago and this month is just an accounting entry that moves it to its final place in the claims account. It's a nonissue in their book. But because it didn't appear in the quality cost report at the time when it *was* an issue, it has totally eluded attention from the quality cost point of view. A big objective in the design of any quality cost system is to be able to strike while the iron is hot. The claim really cannot be booked at the time the reserve is set up. Many will be quick to point out that it isn't a cost yet. The claim may never be paid and almost certainly not in the amount claimed.

One way to deal with this is as follows. Consider that many such claims may be in process at any point in time, some entering the claim reserve, some being flushed out into the claims account. For any month's report the net change in the reserve from the prior period is picked up as a quality cost. This may be a plus or a minus. The input to the claim account is also included. Some will complain that we are entering the claim twice into the quality cost system, once when it is reserved for and once when it is actually booked. But remember, the reserve is a balance sheet account. The claims account appears on the profit-and-loss statement. Consider a scenario, rare in most of these companies, when there is no reserve activity other than one claim. This makes it easier to track what actually happens. In period 1 the reserve is increased in expectation of the payment of the claim. The amount of the increase appears as a quality cost. In period 2 the claim is paid in the exact amount for which it was reserved. The claim account increases by this amount but the reserve *decreases* by this amount. The net effect on the quality cost report is zero. Had we overreserved for the claim, the reserve account would be decreased by more than the actual amount of the claim and quality costs would be decreased by the difference. If we had underreserved, the net effect would be to increase the quality costs in period 2. This treatment goes a long way toward keeping the quality costs contemporary with reality in the minds of management.

For products such as large industrial products, warranty activity is apt to be grossly out of phase with production costs from the standpoint of timing. The following procedure can be used. The quality costs for any given period are never really "closed out" as they would be in an accounting report. Remember, this is an economic model, not an accounting report. Consider a spreadsheet, with months across the top. When required to pay a warranty cost on a product manufactured some months or even years earlier, the operation does not simply book it in the quality cost system as a current-period cost. It is also added to the spreadsheet in the period in which the product completed manufacture. While this is out of phase with the accounting entry, that's all right. It is useful for the custodians of the quality cost system to stress that the consequences being faced today flow from strategies employed some time ago which may or may not still be in place. Management might conclude that the situation was corrected earlier, and by rephasing the costs in the quality cost system they will be able to see the effect of that correction over time. On the other hand, it might alert them to the fact that this may be the first effect of an earlier decision and there is more to come in the future, much more.

Sins We Are Heir To

Ignoring these timing dislocations can be hazardous. Consider a scenario that is hardly rare.

Management by objectives being what it is, a manager of operations that include the quality department as well as manufacturing has been encouraged convincingly to achieve a significant reduction in quality costs. A substantial part of the fiscal year has passed and he has little to show for his efforts, which is fair since his efforts have also been little. If he is to meet his quality cost targets, he must move quickly and certainly. He doesn't necessarily think of it in these terms, but the fact is that there is only one class of quality costs over which he has absolute control and that is the discretionary category. In any event, he decrees a reduction in payroll costs, meaning discretionary costs. As these account for a minority of the total quality costs and most are appraisal costs, he is compelled to take the bulk of the reduction in the appraisal category.

He mandates a severe reduction in the inspection department and at the same time pronounces that manufacturing must assume the responsibility for quality. After all, they make the parts; only they can assure that they are made correctly the first time. The inspectors are dismissed together with a token reduction in the quality engineering department. (There weren't many quality engineers to start with.) Owing to his remarkable grasp of the situation, the quality costs begin an immediate and impressive decline. The discretionary costs are reduced in a step function, as would be expected, but shortly thereafter internal failures begin decreasing! Obviously manufacturing has assumed the responsibility for quality under his sterling leadership. Clearly no one had ever before explained so well the benefits of their participation. He achieves his goal and then some and is enshrined in the pantheon of corporate heros.

Some time not long into the next fiscal year external failure costs begin to increase; modestly at first, then disquietingly, then erupt into a crescendo of warranty claims, product returns, and other forms of disaster.

A postmortem concludes that nothing really changed in the manufacturing processes last year. Those processes were in large part incapable of consistently rendering the products to the requirements that were established for them. With no inspectors to turn to for the close calls for marginal products or to sort the rejected lots, manufacturing supervision took over this duty. This was not an inappropriate action. Of course, manufacturing supervision was always convinced that inspection was fixated on specifications. After all, how could a piece that was 0.0001 inch inside of tolerance be acceptable and of full value while a piece 0.0001 inch outside of tolerance had no value whatsoever? Clearly 0.0002 inch was of no consequence. They began making the kind of "practical" decisions that should have been made all along. After a while they began showing irritation when operators even referred such decisions to them. After all, they expected their people to exercise their own judgment. They were responsible for quality, weren't they? Operators now understood that specifications are really just guidelines. They began exercising their own judgment. Supervisors exhibited vexation when asked to judge parts they deemed obviously usable. They showed even greater irritation when shown parts that obviously had to be scrapped. Therefore, they ceased to be shown many parts.

As operators made increasingly adventurous decisions about the products that they made, the internal failure costs continued to decline. Their competence in making these decisions was supported. Supervision's suspicions about the overly rigorous standards of the former inspectors were vindicated. The manager's leadership in orchestrating this remarkable turnaround was celebrated. Products continued to enter the stream of commerce. They were warehoused, dispatched to mixing warehouses, ordered by customers who put them into their inventories, distributed to stores where they went on the shelf, into homes where they were expected to perform, and into rubbish heaps when they failed to do so. Customers returned them to the stores, stores returned them to their warehouses. Those warehouses were emptied of their unsold product. And demonstrating the American genius, the returned goods were used as landfill under the parking lot for the plant recently expanded to support the booming production. (Author's note: Lest you think this example is hypothetical, be assured that it is not, down to and including the landfill. It would be interesting to hear the speculations of archeologists in future millennia when they discover this treasure trove.)

Summary

Maybe the whole thing started when management looked at quality costs as a direct cost of production to be managed and minimized. While it is all of that, it does not lend itself to superficial management. We can establish an objective to reduce the material bill by 15% next year, for example. We do not, however, just arbitrarily order 15% less material as a forcing function. There is a natural discipline involved here. We must work to change the processes in order to require 15% less material.

We would do better to recognize that quality costs are a trailing economic indicator that give us feedback regarding the effectiveness of our quality improvement efforts. This requires that one accept the concept that quality, in the sense of conformance to the needs and the expectations of the customer, does *not* cost more money. It costs less.

When manufacturers insert highly uniform materials into highly capable processes to render products that external customers value, then they are operating at a low level of quality cost and a high level of profitability.

When quality costs are viewed as a trailing economic indicator in the equation of production economics, then one will *not* point to quality costs and complain that they must be reduced as if they were an independent term in the production equation. Rather, one will point to the level of quality costs and complain that they indicate gross inefficiencies in the production processes and demand to know the sources of those inefficiencies so that they can be eliminated in the future. A subtle, but important, difference.

PART IV

Statistics and Other Laws of Nature

CHAPTER 11

Statistical Thinking

JOURNAL: I think I'm getting the "soft sell." My quality guy keeps talking to me in statistical terms. Not formulas, Greek letters, and that sort of thing, but more like statistical logic, if that makes any sense. Our conversation on inspection efficiency is an example. Next, he's talking in conversational terms about variability, as if I knew what he was talking about. Then almost before I knew it I **did** know what he was talking about. The son of a gun has infected me! I find myself listening to these debates that we have with statistical filters between me and what I'm hearing. When I do that, what I hear doesn't make a lot of sense. One of our guys prefaced an analysis he was about to present with "For the sake of convenience, assume that . . . ." I lost it again. I said, "What you're doing is assuming that the process is predictable and without variation. Yes, that sure **is** convenient! You might just as well assume for convenience' sake that there's no law of gravity and walk out the window at the end of this meeting as assume what you're assuming. Variability is a law of nature, just like the law of gravity. In this case variability is what it's all about. Let's move to the next agenda item. When *you've quantified the variability then come back and we'll talk some more." It was funny actually. Every head in the room, including mine (I'd just heard what I'd said), turned to look at the quality guy. He just shrugged and grinned. Now he's got me doing it. I just might ask our resident wise guy to put together a short seminar on statistical thinking for us. No formulas or Greek letters, just the kind of thing he has done to me. It will be fun to watch him do it to someone else for a change. Most of us had some statistics in college. But somehow we never learned to think statistically.*

We're All Statisticians

It's fair to say that we all think statistically. Every one of us. In fact, it's fair to say that it would be impossible to function in society if we didn't. If we didn't think statistically, we would retire to our beds each evening without the slightest idea of the time to set our alarms for. The only thing that helps us make that decision is the knowledge that tomorrow will be pretty much like today in most respects. That's a statistical notion.

Ask a colleague how long it takes him to get to work in the morning. He will probably (a statistical word) say something like "about (another statistical word) thirty minutes." Now, he may have been making that trip for ten years, five days a week. Press him. "You mean as many times as you have made that trip you don't *know* exactly how long it takes?" Try to ignore it when he looks at you like you're crazy. We all know it's a crazy question. It doesn't take exactly the same time every day and we all know it. That means we understand the nature of variability and central tendency (more statistical language). Your victim knows that most of the trips will be very close to thirty minutes but most will vary a bit over or under.

Press on! Ask your colleague to estimate the variation from trip to trip. What is the "time window" that will include most trips? At first he might say, "Plus or minus ten minutes," furtively looking about to see if anyone is listening to this exchange. Reinforce your obsessive interest by boring in further. "You mean it's just as apt to take you twenty minutes as forty?" "Well, not actually. Twenty-five minutes is about the best that I will do, but it can take as much as forty-five. But most will be about thirty. That's my average." This genius understands the concept of variation, central tendency, can estimate variation, and is familiar with the shape of the distribution curve. You should be flabbergasted at his command of statistics. After all, you are talking to the elevator operator! This is pretty arcane knowledge for an elevator operator!

Now let's really get sophisticated. Ask the fellow why the trip can vary as much as twenty minutes from one day to another. "Well, it's not too often as little as twenty-five minutes as much as forty-five. Just occasionally. But why? Well, one factor is how the stoplights work out; some trips are just better than others. If it's a nice day, I just may relax a bit and drive a bit slower. There are a lot of reasons, but they all average out for the most part." So the man understands the behavior of averages and how normal variation works.

Then ask, "What would your boss say if you were late and you told him that your trip to work took an hour and a half this morning?" "He would probably ask, 'What happened?' " So performance out of the expected range is automatically attributed to some out-of-the-ordinary occurrence that can be identified. It's not often that all of those little things pile up on one side of the equation. They tend to average out, as you said. "And can you explain to him what happened?" "Of course. If that happened, there would surely be an obvious reason that I could offer." This person even understands the concept of random variation versus assignable cause.

"When do you leave for work in the morning?" "About forty-five minutes before my shift begins." "Why? It only takes you thirty minutes to get to work?" "Sometimes it takes longer and I don't want to be late." "What would you do if some morning it was absolutely essential that you be here on time?" "I'd probably leave an hour early." This person even understands confidence and how to manage it.

These are all statistical concepts and some of them are pretty sophisticated. Yet as you read through this scenario, it probably (that statistical word again) sounded totally ridiculous. Everyone knows these things. If you didn't believe in the notion of central tendency, then the fact that it has taken you about thirty minutes to get to work for the last ten years wouldn't mean a thing when it came to predicting tomorrow's trip. As a result you wouldn't know how long the trip might take and therefore what time to set your alarm clock for.

We *do* understand statistical concepts. Our language is full of statistical references. Words like "probably" and "about," expressions like "it all averages out" and "I'll take a chance" imply statistical concepts. The only difference between a statistician and a layman on the subject is that the statistician has reduced these natural probabilistic tendencies to mathematical terms. He can process data to yield estimates that are quantified and can tell you in numerical terms what confidence (or its opposite, risk) is attached to his estimates. This is not a trivial skill. But it is nonetheless based on natural laws that are so ingrained in all of us that we don't even appreciate the fact that we possess the knowledge.

Characteristics of a Distribution

There are three characteristics of a population that, taken together, tell us about all there is to know about it.

Central Tendency

One is its central tendency. That is, what is the value about which the distribution is centered? This is defined as the mean (average), the median (the central value if you were to put all the numbers in ascending or decending order), or the mode (that number or cell that occurs most frequently). The mean is the most popular of the three, but there's a problem with the use of the mean. It doesn't necessarily exist in reality! Consider the die (that's half a pair of dice). What is its average value? Careful now, think before you answer. It's $1 + 2 + 3 + 4 + 5 + 6/6 = 21/6 = 3\frac{1}{2}$. Has anyone ever seen a $3\frac{1}{2}$ on a die? Funny, when you consider that the most frequent outcome when rolling two dice is seven, or an average of three and a half each. More on that later.

Variability

The second characteristic is variability. How much do the individuals in the population vary about the central value, the average for instance? The most common measure of variability, or dispersion if you prefer, is the "standard deviation." While this term has a certain metaphysical ring to it, like many terms created by statisticians, there is nothing particularly mysterious about it. After all, it took a statistician to articulate the concept of the "normal deviate." In fact, the term is so straightforward that many of those who use it regularly have a difficult time explaining it to those who don't. Let's look at a mathematical expression, the formula for the standard deviation of a sample taken from a population.

$$S = \sqrt{\frac{\Sigma(x-\bar{x})^2}{n-1}}$$

where

    S = standard deviation of a sample

    Σ = sum of

    x = an individual value

    $\bar{x}$ = the average of all the values

    n = the number of values used in the calculation

The expression $(x - \bar{x})$, then, is the difference between each reading and the average of all readings, a deviation in other words. We square each, giving us the square of the deviation. We then add all these squared deviations and divide them by the number of deviations in the sum. This gives us the average, or mean, of the squared deviations. Having squared the deviations before we averaged them, we now extract the square root of this average, which "sort of" returns us to our original terms, but not quite. What we have at this point is the square root of the mean of the deviations squared. The shorthand for this term is "root mean squared" (RMS) deviation.

Now, the expression "RMS" is probably familiar to many of us even though we have never been quite sure what it means. Engineers speak of surface finish of machined parts in terms of RMS value. This would be a measure of this type of average deviation of the surface from the average of the whole surface, in other words roughness. Electrical engineers speak of RMS voltage in certain circumstances.

Why go to all this trouble? Why not just operate on average deviation? Well, we could and sometimes we might. The parameter called "variance" is standard deviation squared, or the above expression before you extract the square root of the average. It has some useful properties and is much used by those trained in statistical processes. But standard deviation also has some interesting properties. Consider the following example.

Earlier standard deviation was referred to as a "sorta" average. What does that mean? It means that this calculation gives a great deal of weight to large deviations from the average and actually suppresses small ones. Consider several deviations going into the calculation of the average under the square root sign. Let's say that the value we have measured is thirty and the average is twenty. The deviation is ten, and its square that we put into the calculation of the average becomes 100! The next value is ten. The deviation, therefore, is minus ten. When we square that we get 100 again, $-10 \times -10 = 100$. Therefore, a negative deviation doesn't cancel out a positive deviation when we square them. We are interested in the tendency to deviate, not the direction. A deviation is a deviation. This seems to make sense so far. The next value is twenty-one. The deviation is one. When we square one, we get one! The last two deviations we got we increased by ten times before we put them into the average deviation; this time we didn't multiply it at all. The next value is twenty and a half. The deviation is one-half unit. When we square this deviation, we get one-quarter, $\frac{1}{2} \times \frac{1}{2} = \frac{1}{4}$. The number that we put into the average deviation actually got smaller. So in calculating standard deviation we penalize large deviations. The penalty gets smaller as the deviation gets smaller, until tiny

.eviations are actually forgiven to a degree. Therefore, standard deviation is sort of a weighted average deviation. The author realizes this hasn't been a particularly elegant explanation. But maybe it has helped demystify the subject. Besides, it's all an estimate anyway.

A note of caution. The casual reader is not necessarily equipped, based on this discussion, to venture too boldly into the world of statistical analysis. Hopefully, he *is* prepared to listen to the cognoscenti with a bit more discernment.

Distribution Shape

The third parameter is the shape of the distribution. The most frequently mentioned is Gaussian distribution. You've never heard of it? It's the same as the normal curve or the bell-shaped curve. It's "The Curve," that mysterious standard against which we were measured in school, usually by someone who didn't have the vaguest idea of statistics or how to use "The Curve."

The expression "normal curve" is unfortunate. It suggests that all other distribution forms are "abnormal," somehow not quite right. This isn't true at all. Roll a die a million or so times, plot the outcomes, and you will get a rectangular distribution, about as many of one number as of any of the others. In fact, if you don't, you've just rolled a crooked die a million times.

The normal curve does, however, appear so frequently in nature that we begin to expect to see it. In many circumstances if the data doesn't form a bell-shaped distribution, we become instantly suspicious of the data.

Also, it seems almost as though distributions sometimes conspire to deceive us as to their shape. This is particularly true when we consider samples taken from populations. A statistical control chart is a series of samples taken from a population. Some pretty heavy decisions are made based on control charts and it pays to be sensitive to the ways in which these statistics deceive the unwary. Whoever said that numbers don't lie never dealt with sample statistics.

Roll the die again a very large number of times, plotting each outcome. Better than that, let's reason our way through the process. We will assume for the purpose of developing this model that the die is "honest," meaning that on any roll each of the six possible outcomes is equally likely to occur. Over time we would expect to get an equal number of each outcome (theoretically)—that is, a rectangular distribution as shown in Figure 11.1.

Now take two dice. But instead of dealing with the sum of the two, let's plot the average of the two dice. Six outcomes are possible on each die, a total of thirty-six. These are enumerated in Figure 11.2.

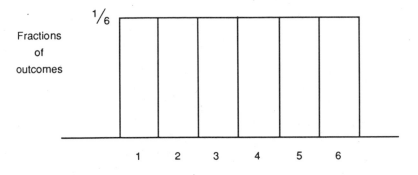

Figure 11.1. *Theoretical distribution of outcomes of the roll of one die.*

Die 1

		1	2	3	4	5	6
	1	1.0	1.5	2.0	2.5	3.0	3.5
	2	1.5	2.0	2.5	3.0	3.5	4.0
Die 2	3	2.0	2.5	3.0	3.5	4.0	4.5
	4	2.5	3.0	3.5	4.0	4.5	5.0
	5	3.0	3.5	4.0	4.5	5.0	5.5
	6	3.5	4.0	4.5	5.0	5.5	6.0

Figure 11.2 *Possible outcomes of the average of two dice.*

How many of these outcomes will average one? Obviously, one. It takes a one on each die. What about two? Well, there are three ways, a one and a three, a three and a one, or two two's. An average of three can occur five ways. Four can also occur five ways. An average of five can happen three ways and six only one way. Add all these outcomes and we have accounted for eighteen outcomes of the thirty-six. What happened to the missing outcomes? Since we are using two dice we can get half values. For example, we can average one and a half two ways: by rolling a one and a two or a two and a one. Plot the distribution of the thirty-six outcomes on a blank piece of paper. It should look like Figure 11.3. But look at the

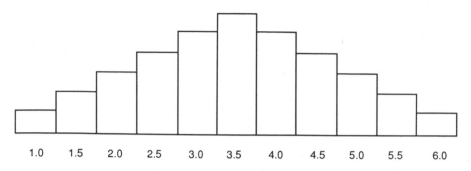

1.0   1.5   2.0   2.5   3.0   3.5   4.0   4.5   5.0   5.5   6.0

Figure 11.3.   *Theoretical distribution of outcomes of the average of two dice.*

shape! We have taken sample averages of two from a known and well-understood rectangular population and the shape they form is triangular, not rectangular!

Now take samples of four dice and plot the averages. There are 1296 possible outcomes. Four dice, each with six possible outcomes, equals 6 × 6 × 6 × 6 = 1296. How many of these average one? One, that's all. Also, one outcome averages six. But the averages can now take on values of one-fourth because we have four dice. Enumerate all the possible outcomes and plot the results and you will find that the shape of the distribution of sample averages taken from the rectangular population will be closer to the shape of the classic normal distribution than anything you are apt to find in nature. If we plotted a large number of samples of four, we would be absolutely convinced that a die reflects a normal distribution! A distribution of sample averages of four or so will *always* assume a near-normal shape *regardless of the shape of the parent distribution!*

Nonetheless, many are fooled into believing that they are dealing with normally distributed characteristics with processes with which they are not so familiar. Figure 11.4 shows all three of these distributions superimposed on one another. Let's see what we can infer about sampling from these curves. All are symmetrical and average 3½. This is the true average even though it is impossible to roll a 3½ on a single die. Averages don't have to exist in reality.

As sample size increases (the number of dice used in each roll), the error of each individual trial from the average decreases. That is, the tendency is for the outcome of the trials to be closer to the true average. Let's examine that. If we didn't know and couldn't reason the model of outcomes for a die, we would have to approach the issue experimentally. An individual

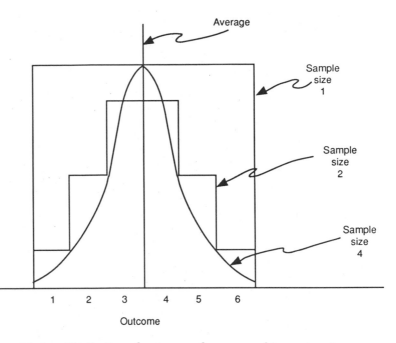

Figure 11.4.   *Distribution of outcomes of averages of increasing size.*

roll of one die is as apt to render a one as it is a six, both the maximum error possible in suggesting the average. Moreover, it is impossible for one roll to tell us the true average (3½).

When we increase the size of the samples used from one to two, it is possible for the outcome of an individual experiment to tell us the true average (3½). In fact, three and one-half will be the most frequently occurring outcome when the experiment is replicated a number of times.

By simply increasing the sample size from one to two in each experiment the average of the two will not only tend to the truth, but the occurrence of the maximum error diminishes radically! With one die a one or a six will occur one-sixth of the time each. With two dice these extreme outcomes will only occur ⅓₆ of the time each. The outcomes begin to crowd in around the true average. By increasing the sample size we have constrained the ability of the experiments to lie to us. We have, in the words of a statistician, reduced the "error of the estimate."

When we increase the sample size in the experiment to four, the results of the individual experiments crowd in further around the truth and large errors become even more unlikely. An average of one or six will occur only one time each out of 1296 trials.

Note that an arithmetic increase in sample size gives us a geometric reduction in the error of the estimate. The denominator of our error term went from 6 to 36 to 1296 as our sample size went from 1 to 2 to 4. That's efficiency! One might even note that the denominator in each case was the number of possible outcomes on an individual die raised to the power of the sample size: 6^1, 6^2, 6^4. This phenomenon underlies lot and process sampling (control charts).

CHAPTER 12

Statistical Process Control

JOURNAL: That session we had on statistical thinking helped a lot. I sense that some of the functional executives are thinking differently about data. More important, they're asking different kinds of questions when data is presented to them. They're relating better to some of the concerns of the quality organization. It's said that the common language of business is dollars. Perhaps the common language should be statistical thinking. So many things can be discussed using common statistical terms.

I've gotten more interested is statistical process control too. I think there are probably applications outside the factory. The quality guy explained control charts and how they work. They're so straightforward when they're explained right. Why do we make simple things seem so complicated? Of course, Frank has a knack for taking the complicated and making it simple. That's a real gift.

Wanting to get some ''hands on'' experience with control charts, I remembered that I had a ready-made opportunity, the daily weight chart I keep on the back of my closet door. Taking the last sixty days or so, I converted the data to a control chart. I got my first real lesson on the difference between process control and process capability. I've been a bit fuzzy on the distinction until now. At breakfast the next morning I proclaimed to Mrs. CEO that my weight was in excellent statistical control. It averages 170 and my control limits are 158 and 182. I've not exceeded these limits in sixty days. It was then I remembered where I studied Executive Eye Contact 101. Mrs. CEO drew a bead on me and said, "I don't know about this control business but I do know that you weigh too much. One seventy is unhealthy. At 182 you're a real porker and the last time you saw 158 was when we came back from Mexico where you had a week-long case of Montezuma's Revenge." I remembered a comment by Frank about a process of ours. I thought he was being sarcastic when he commented that it was in near-perfect statistical control, making almost 100% defective product. My weight is in perfect statistical control, meeting neither my needs nor my wife's expectations. She says that I should be at 165 and not exceed 170. Less than 160 and I look gaunt. Over the months it's become easy to talk to Frank, so I shared this with him. Apparently wives are no respecters of rank. He has had an identical experience. He noted that one's weight is an excellent case study, pointing out that I had two issues with which to deal. Average and variation. Frank noted correctly that I knew how to deal with the average. Enough said. As Joe Juran, the eminent quality guru, might say, it's a problem not a fate. As with a lot of industrial processes, we know very well how to control the average, we simply don't choose to.

Variability is a bit more complicated. There are two approaches, it seems to me. If the objective was to not exceed 170, I could bias the average down to 158, and if the variation remained the same, I would be okay. Of course, my resident chief engineer doesn't want me below 160, so scratch that. My job is to reduce the variability. This was instructive. I took my chart in and showed it to Frank. (Threatening his life if word got out that we were sitting around doing statistical analysis on my weight.) He showed me where the pattern didn't look right. Not enough runs above and below the center line. It suggested to him some nonrandomness. He scribbled a few numbers and proclaimed that my Monday weights were all above the average and all my Friday weights were below. In fact, my lowest Monday weight is higher than my highest Friday weight. There was evidence of a "statistically significant difference" between my Monday weights and my Friday weights. The good news is that the variation among Fridays is fairly small. So is the variation for Mondays. It appears that the excessive variation is due to the weekends. I do look at weekends as a vacation from discipline, spending three days of the next week getting back down. More day-to-day variation. Then something he once said occurred to me: "The operator can usually control process *centering if he is inclined to. Reducing variation, however, usually requires us to redesign the process in some way.''* Well, I'll be!

We the Jury Find the Defendant . . .

The majority of all statistical techniques that one comes in contact with, in manufacturing particularly, are tests of hypothesis. Consider a process average. This is an important parameter of a production process. We know that it is constantly varying, hopefully by an insignificant amount, but constantly varying nonetheless. That's nature. At the risk of redundancy, we can say that the process average is constantly "hunting" or varying about the long-term average. If we didn't believe this, we wouldn't be sampling the process. This fact, combined with what we know about the error of the estimate, leads us to conclude that one cannot with certainty determine what the process average is at any moment in time. However we can determine with reasonable precision what it is *not*. Therefore, we establish a hypothesis that we will accept, barring substantial evidence to the contrary. We then design an experiment to collect data with which to test the hypothesis. This really isn't as difficult as it sounds. We have another model.

Consider the American criminal justice system. It works exactly the same way. The system has a hypothesis that it will accept barring substantial evidence to the contrary: The defendant is innocent. Evidence is collected and presented to the jury. The jury is instructed that they must find the defendant guilty *beyond reasonable doubt* before they can render a guilty verdict. Also, note how a favorable verdict is expressed. The jury doesn't say, "We find the defendant innocent." The defendant is *presumed* innocent. Rather they find the defendant *not guilty*, meaning that the evidence presented does not constitute evidence beyond reasonable doubt of guilt, and therefore innocence is presumed. We know that the defendant may indeed be guilty, as sin; or guilty of sin to be more exact. But our system is biased in favor of the defendant. Does the system have to work that way? Clearly not. History is full of systems that are biased the other way. They presume guilt, not innocence. Proving innocence is often impossible, statistically as well as legally.

Statistical process control works the same way. The process is presumed innocent of misbehavior, barring evidence beyond reasonable doubt to the contrary. In the absence of such evidence we keep on producing. The process may be guilty as sin, but we haven't proven it, yet. But at least the statistician can quantify what is meant by "reasonable doubt" in probability terms. That is more than a jury can do.

This is a very important point. Beware the person who points to a control chart and says, "See, the process average is right where it's supposed to be." The more careful observer will say, "See, we have no evidence that the process average has moved."

Control Chart for Averages

Consider that we have a process for which we wish to institute control charting for averages. We have taken a reasonably large sample size of *individual* values and plotted them. The distribution has a shape reasonably approximating the normal curve.

We're not too fussy here usually. "Reasonable" is a very elastic word. Nature rarely, if ever, shows us a perfect, or even near-perfect, normal distribution. But we do know, don't we, that we can't infer the shape of the distribution of individuals by looking at a distribution of sample averages. Sample averages of as few as four will produce a distribution closer to normal than almost anything nature will give us—regardless of the shape of the parent distribution.

Consider that we intend to take samples of four with which to operate our control charts. We also know that as the sample size increases, the variation of the sample averages decreases from that of the individuals of the parent population. We demonstrated that with dice. What do we do with this problem? After all, we're not selling process averages to our customers, we're selling individual items. Fortunately, this is not serious because we know a great deal about the behavior of samples. Specifically:

$$\sigma_{\bar{x}} = \frac{\sigma_x}{n}$$

where

$\sigma_{\bar{x}}$ = standard deviation of sample averages

σ_x = standard deviation of individuals

n = sample size

Now to the casual observer this may not look like the vital information, but it is. It says that the standard deviation of sample averages is related to the standard deviation of individuals from a normally distributed population.

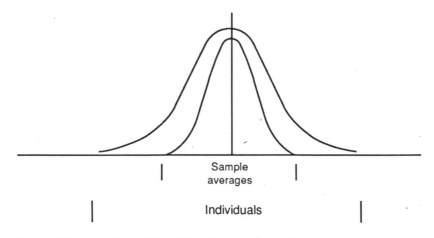

Sample
averages

Individuals

Figure 12.1. *Relationship of distribution of sample averages to individuals.*
Sample size = 4.

That relationship is determined by the square root of the sample size. Therefore, the standard deviation of sample averages of four taken from a population will be half (square root of 4) of that of the individuals. If we're not sensitive to this point, we're apt to compare the statistical limits of a control chart to the limits of a specification. Finding the distribution of averages somewhat inside the specification, we are reassured. But the fact is that the distribution of individuals is greater than the sample averages and we may be producing substantial quantities of nonconforming product. This phenomenon may explain why so many people are convinced that statistical tools don't really work. The truth may be that these people don't know how to use them.

Let's build a control chart. The large number of individual values has been graphed in Figure 12.2. Consider this a population from which we will sample. We'll treat it as normal because we have judged that it is "close enough." That, also, is a statistical concept. Were we to take a large number of samples of four from that group (throwing them back in and shaking the bag each time to keep from depleting our population), we know that we would get a normal distribution one-half the width of the distribution of individuals shown in Figure 12.2.

As long as the variability of the process (the parent population) and its average remains unchanged, we know that samples will fall within the sampling distribution that we have determined (with the occasional and rare exception; life is never without some risk). We could at this point

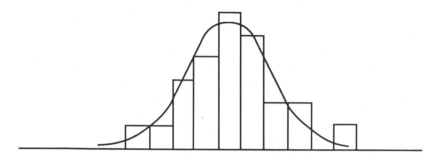

Figure 12.2. *Population distribution.*

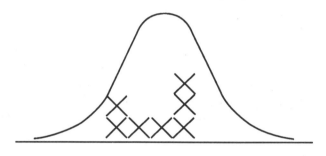

Figure 12.3. *Sampling tally sheet.*

simply start sampling and putting X's on the graph as shown in Figure 12.3.

But that's not totally satisfactory. We would quickly run out of graph, and besides we would like to see the picture unfold before us. There may be a trend developing and the graph shown in Figure 12.3 won't reveal it very well. Why don't we turn it on edge and extend the average and the expected limits of the distribution of sample averages so that we can plot the time series? That would look like Figure 12.4.

We now have a control chart. The control limits on a control chart for averages are the three standard deviation limits of samples of known and constant size taken from a statistically described population. Because the normal distribution is described mathematically, the plus and minus three standard deviation limits include about 99.7% of the total population. Yes, there will be the occasional outlier.

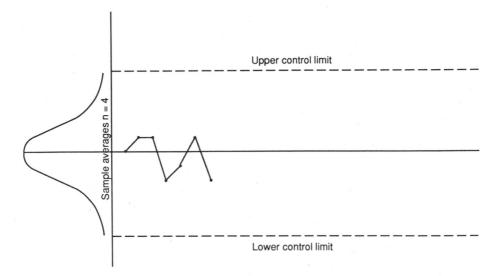

Figure 12.4.   *Statistical control chart for averages. n = 4.*

We can now say that a point outside the control limits constitutes evidence beyond reasonable doubt that the process average has changed, for the simple reason that if the process average *hasn't* changed there are only about three chances out of a thousand that it will produce a point outside these limits. Another way to say it is that the process average *has* changed at a risk of three chances out of a thousand of being wrong. That constitutes evidence beyond reasonable doubt that the process average has shifted.

This, then, is a commonly accepted rule for interpreting a control chart: one point out of the upper or the lower control limit. There are others.

Extended Rules

The remaining rules are referred to as "extended rules." That's an unfortunate choice of words. It implies to many that these rules are extrapolations on the basic rule and somehow not quite as legitimate or at least not as important. Nothing could be further from the truth. These additional rules come from application of the same rules of probability and are just as legitimate as the "single point out" rule. In fact, there is probably more to be gained in using the extended rules than the other.

For a single point to be out of the control limits takes quite a move in the process average relative to its normal behavior. If the process was set up properly in the first place, a jump of three standard deviations is a lurching change. A fixture slipped, a cutter broke, a timer was bumped, that sort of thing. It is more common for processes to drift or to be set up just a bit off. When the "extended rules" are ignored for the sake of simplicity of training and administration, we are usually throwing out an enormous amount of meat with the bone. What if the cutter slipped, but just a little bit. This will never be picked up using the three-standard-deviation rule alone.

Often the extended rules are written down for the benefit of the operators. This is fine. But it does suggest that there are a small number of extended rules. They also sound rather arbitrary, which doesn't help; "seven consecutive points on one side of the center line" and so forth. Actually, there are a near-infinite number of "extended rules" and there is nothing arbitrary about them.

One point outside the control limits can happen with an "in control" process about three times out of a thousand. Apparently that's the risk of a wrong decision that we are willing to accept. There are other combinations of outcomes that would allow us to claim that the process has shifted with similar risks.

Assume that the process in question is operating exactly on its long-term process average (the center line). Remember, this is the base assumption, also known as the null hypothesis. The process is innocent of misbehavior barring substantial evidence to the contrary. What is the probability that the next sample will be above the center line? Right, 50%. Obviously there is also a 50% probability that it will be below. What is the probability that the next two points will be on the same side of the center line? Remember the null hypothesis. Each individual probability is 0.5 (50% if you prefer). Since each outcome is independent of the other, the joint probability must be $0.5 \times 0.5 = 0.25$, or 25%. Could this happen with a process that is "innocent"? Yes, one time in four. What about five in a row? The probability there is $0.5 \times 0.5 \times 0.5 \times 0.5 \times 0.5 = 0.03125$, or 3%. Let's try seven. By now you realize that we can express this as $0.5^7 = 0.0078125$, or about 0.7%. In a like manner it is possible to calculate a large number of combinations of outcomes that one could use for decision criteria.

Control Chart for Ranges

The control chart for averages is usually accompanied by a control chart for ranges. This chart operates off of the within-sample variation, the range

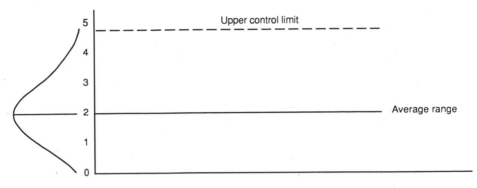

Figure 12.5. *Distribution of sample ranges having an average of two.*

between the largest and the smallest value in the sample. It follows that the range is always a positive number. The distribution is not symmetrical, as in the case of the chart for sample averages. In a sample of four, for example, it is entirely possible for the range within the sample to be zero, that is, four identical readings. The average range might be quite small, not much above zero. But the maximum range that could occur with reasonable probability could be quite a bit larger. As a result, the distribution of sample ranges will have a "tail" to the high side. For reasons beyond the scope of this book, the lower control limit on range when using sample sizes of seven or fewer (there are reasons for not using more) is always zero. Using a sample size of four as an example, if the average range were to be two, the lower control limit on range would be zero and the upper control limit would be 4.56. This is shown in Figure 12.5.

Other Forms of Control Charts

As control charts for sample averages and range are the straightforward products of applying the laws of probability, it follows that we could create other types of control charts.

Attribute Control Charts

Attribute control charts are often disparaged as being crude and unsophisticated. This is unfortunate. For one thing, what if the characteristic being dealt with is an attribute? Try as we may, we're going to have a difficult time extracting variables data for an attribute characteristic. Also, what if we had ten characteristics, all variables and all important, to be monitored.

This would require ten control charts for averages and ten for ranges, a near-impossible number to be managed. But although we cannot treat an attribute as though it is a variable, we can treat a variable as though it is an attribute. We do this every time we choose to use functional or go-no-go gaging. Under these circumstances we might consider setting up one attribute chart to control all ten characteristics using go-no-go gaging to justify the larger sample sizes needed. We *would* be well advised to accompany the control chart with a way of logging any defects that are detected by type.

Types of Attribute Charts

There is also a variety of types of attribute control charts. One is the percent defective chart. In this case the percent defective in the sample is calculated and plotted. If percent defective can be used as the basis of a control chart, then it follows that the number of defectives can be used, saving the operator a little arithmetic. Let's look at that for a moment. If percent defective is the number of defectives divided by the sample size, and the sample size is constant, then it follows that we could accomplish the same end by just plotting the number of defectives. Why ask the operator to perform a calculation on every sample just to put the statistic into terms that management can easily relate to? It seems as though it should be the other way around.

Instead of dealing with defectives (number of units bearing one or more defects), it may be beneficial on occasion to deal with number of defects in the sample. Consider a complex product, final inspection of an aircraft carrier, for instance. I suspect that under rigorous inspection there is no such thing as a defect-free aircraft carrier. Using a percent defective chart on our aircraft carrier production line will show a process average of 100% defective. However, if we decided to count the number of defects encountered, we would have a useful statistic. We could track the total number and hopefully watch it decline over time.

Still No Free Lunch

For all their charm, and sometimes necessity, there are a few drawbacks to attribute control charts, For one, the sample sizes required are large, as in all attribute sampling plans. Much of this disadvantage can be brought back when functional gaging is substituted for attribute gaging.

A more frustrating issue is the fact that as the quality gets better, the sample size must be increased! This is contrary to intuition. It is also contrary to the natural tendencies of management to spend increasing amounts of money to assure a process of increasing performance. But consider that the sample size must be large enough for the tendency of the process to produce defects or defectives to manifest itself in the sample. Without getting into a lot of arithmetic, let's reason our way through it.

Using a percent defective chart, let's consider that the percent defective of the process is quite high, say 10%. A sample size of thirty pieces would be expected to contain, on average, three defectives. The upper control limit would be higher yet. With this sample size we could "see inside" the process at work and monitor its performance short of producing an unusual number of defectives. In other words, we could observe improving or deteriorating trends of defectives without exceeding the upper control limit.

If the process were to be improved to the point where it is producing 0.1% defective, we would be in trouble with this sample size. On average we would expect to see 0.03 defective per sample. The upper control limit would be less than one defective. The process would either be "good" at zero percent defective or "bad" at one or more defectives. But one defective could represent quite an increase in the process average. To get insight from a process sampling plan like this, it should be possible for the process to produce one or more defectives in the sample without being called out of control. In short, the sampling plan has lost sensitivity relative to the quality being produced. To gain that sensitivity back it is necessary to increase the sample size.

Other Charts for Variation

Suffice it to say that other charts can be developed to deal with variability besides the range chart. Control charts can be developed for moving range, standard deviation, or virtually any other measure of variation that one might choose to operate on.

Worship of False Gods

One question we should ask ourselves is "What's so magic about three standard deviation limits on control charts?" Nothing, as this author sees it. If this point of view has any validity, then why do we invariably assume three sigma limits on our charts? Most control chart forms have tables and formulas on the back to help in setting the limits. These factors, while very helpful, always presume three standard deviation limits. In fairness, the reason is to maintain the maximum amount of simplicity for the practitioner. Operators can't be expected to be experts in statistical theory. But typically, operators don't set up their own control charts; engineers do. They *can* be expected to be expert in the use of control charts.

If I am operating a process that is easily adjusted, the turn of a knob for instance, and the consequences of misadjustment are minimum, do I *really* want to refrain from adjusting the process until I have reduced the risk of misadjustment to two or three parts per thousand? This is conservative, yes. Statisticians are understandably conservative people. But managers make decisions every day risking huge amounts of money with greater chance than two or three parts per thousand of being wrong. Perhaps the question should be "What is a prudent amount of risk to accept in making this decision to investigate or adjust the process?" Then we should set up our control charts accordingly.

The Bottom Line

The bottom line is that we should be as familiar with the laws of probability and the concepts underlying common statistical tools as we are with the law of gravity. Both control our behavior, whether we like it or not. Few of us can calculate offhand the rate of acceleration of a free falling body or the velocity upon impact following a fall of 200 feet. "Academic," we might say. But we still know that there is a difference between jumping off a chair and jumping off a building. We have a conceptual understanding of the behavior of gravity and how it influences the decisions we make.

We should have a similar conceptual understanding of the laws of probability and how they affect the decisions we make every day. Yes, others can calculate the probabilities and work out the risks to many decimal points. That's useful and necessary. But we should all consciously understand the concepts of applying these natural laws to our everyday lives.

Journal

Well, another year has passed, and quite a year at that. Rather chaotic in many respects. But it's been a good year, even though we haven't much more than gotten started.

What have I learned? For one thing, I've learned how naïve I was! I knew a big change was needed. I knew it would take a lot of effort to make the change. But I just didn't have any idea of how stubborn and intractable are the ways of people. Or maybe I just overestimated my power as "the boss."

*I've sure learned that the change has to start with me. First, in my head; then, in my gut. I wasted a lot of time trying to convince others to change. Once I realized that I had to change, **thinking** and **behaving** my way through that change very deliberately, I noticed that a few others began to follow.*

I've learned that how a leader talks doesn't count for much. It's how he behaves. People believe what they see, not what they hear. The message to me and my management team is "Don't tell me what you believe—show me!"

I've learned that this kind of change is a product of stubborn, single-minded leadership; certainly not management of the plan, organize, staff, and control type.

What have we accomplished this year? Quite a bit actually. First of all, I've changed. And I've crystallized a lot of anxieties and fuzzy feelings into something concrete that I can wrap my mind around and get passionate about in a very unequivocal way. I know where we have to go and I know we can get there.

I've got a vision for this company now. That was tough. People like the vision. They feel good about it. They can relate to it and buy into it. One crusty old goat (about my age) in one of our factories put it very well during one of the skip level meetings where I was trying out some thoughts. I was trying to describe our company as I see it. "Sounds great," he said, "I'd like to work for a company like that someday." The room went quiet.

People were waiting for me to go ballistic. I couldn't help it. I laughed harder than I had all year. When I got myself together, all I could say was "So would I, why don't we do it here?"

We're starting to manage a bit more on the basis of fact rather than assumption and folklore. That's been tougher than I would ever have believed. We assume that managers and engineers are compulsively objective and fact based. But then we assume a lot of things, don't we?

Our priorities make sense to most people at all levels. That has helped us to keep focus. Increasingly those priorities and values are echoing through the organization's decisions.

People are beginning to get the hang of the value of quality cost information. Their analyses are less academic. They're beginning to use the information in making their decisions and plotting their strategies.

Continuous improvement. That's where it's at. No doubt about it. We still seem to be driven by emergencies, the fire drills when customers reject our products, that sort of thing. But I'm beginning to see some evidence of continuous improvement in action.

I've got to get us to the point where process improvement is a hero maker as much as problem solving. I've also got to break down the fiefdoms that get in the way of real progress.

This coming year we have to get at the bottom line. Quality has to go up and costs have to go down in a significant way. We've got to find out more explicitly what our customers expect and what they like and don't like about our products and services vis-à-vis our competitors.

We've got to learn how to get new products to market faster without sacrificing quality.

We have to create some new reward systems. Our managers have to learn that they're not going to prosper in this organization unless they deliver objective evidence of continuous improvement. It's time to set out some truly challenging targets for the organization in ways that can't be ignored.

This year we must institute the use of statistical tools at the top of our organizations. Unless managers and executives are using these tools to manage and improve their processes, we can't expect them to be used by the rank and file. When managers get used to using them and develop an appreciation for their value, they will begin to think statistically. Only then will they begin to ask statistical kinds of questions of their people. Until that happens statistical process control will remain an aberration; something off to the side to please the old man in his dotage. SPC is the key to consistency. We must not only improve quality and reduce cost, but we must improve consistency of all our processes.

Education and training. Again, two words for the same thing, I used to think. Not so. Training has to do with imparting vocational skills. Education has to do with changing the way that people think about things. We've wasted a lot of time and money training people this year. There, I said it. Pure blasphemy! But, we've poured training, good training, all over people who never had any intention of using it on the job. Some saw their training in new skills as an effort by management to get more out of them without paying them for it. Others simply failed to see the relevance of what they were being taught to the performance of their jobs. The implementation of statistical process control is a prime example. We have more control charts and less statistical process control than I would have imagined possible! Somehow we must do a better job of mind preparation before we start imparting new skills. That's education.

Yes, I'm looking forward to the coming year. One thing for sure. It won't be like the year behind us. The issues are different. We've got to produce those results I was looking for last year, and not in a token way.

What was it the quality guy said during one of our talks last year? "In my business we deal mostly in small victories." That's true of my business too. From that point of view this year has been a success. We have had our share of victories, and not so small at that. A year ago I was looking for the bases-clearing home run. But first we had to put men on base. This year we did that. This year had to happen before next year could be possible.

QUALITY IS A JOURNEY, NOT A DESTINATION. . . .

SURRENDER? . . . NUTS!

Appendix: Definitions

Culture A community's response to its environment.

Inspection effectiveness The fraction of the defects or defective products that inspection correctly identifies and removes.

Inspection efficiency The percentage of inspection decisions correctly made.

Leadership The exercise of moral authority that causes others to willingly follow, sacrifice, or assume risk.

Management The process of allocating scarce resources to the achievement of enterprise objectives.

Moral authority Authority conferred from below based on performance and personally earned respect.

Paradigm A model of reality.

Process capability The relationship of the spread of a process operating in a state of statistical control to the spread allowed by specification or regulation. A process is judged capable when its spread fits within the allowed limits.

Process in control A process is judged to be in control (statistical control) when it is operating within a system of constant causes and therefore is predictable within calculable limits.

Quality The degree to which products and services satisfy the needs and expectations of the customer and the ultimate consumer.

Rectifying inspection *See* Remedial inspection.

Remedial inspection Inspection conducted for the express purpose of improving the quality of the product stream exiting the process (weeding out defective products).

Statutory authority Authority conferred by rule, regulation, or law and enforced with the power to sanction or reward.

Submitted quality level The percent defective or defects per 100 units present in the production submitted to an inspection process.

Verifying inspection Inspection conducted to confirm the continued satisfactory operation of a process.

Waste Cost or time needlessly consumed in activities undertaken to meet the goals of the organization.

Cost Terminology

Administration costs The cost of those activities essential to the effective administration and management of the quality program which cannot realistically be categorized as either preventive or appraisal.

Appraisal costs The cost of those activities carried out to assess the quality of the products of the organization.

External failure costs The cost of correcting or replacing products improperly rendered and which were detected after delivery into the distribution system.

Internal failure costs The cost of correcting or replacing products that have been improperly rendered and have been detected prior to delivery into the distribution system.

Prevention costs The cost of those activities carried out with the objective of avoiding the creation of nonconforming products or events.

Process scrap costs Scrap or waste inherent in the operation of processes using the current methods, materials, and equipment.

Index